Thomson Nelson Guide to
Success in Social Science

Writing Papers and Exams

Diane Symbaluk

Thomson Nelson Guide to Success in Social Science:
Writing Papers and Exams

by Diane Symbaluk

**Associate Vice President,
Editorial Director:**
Evelyn Veitch

Acquisitions Editor:
Cara Yarzab

Marketing Manager:
Laura Armstrong

Developmental Editor:
Sandy Matos

Production Editor:
Tannys Williams

Copy Editor:
Maya Bahar

Proofreader:
Nancy Mucklow

Senior Production Coordinator:
Hedy Sellers

Design Director:
Ken Phipps

Interior Design:
Katherine Strain

Cover Design:
Courtney Hellam

Cover Image:
Paul Thomas/The Image Bank/
Getty Images

Compositor:
Courtney Hellam

Printer:
Webcom

**Library and Archives Canada
Cataloguing in Publication**

Symbaluk, Diane, 1967–
    Thomson Nelson guide
to success in social science :
writing papers and exams /
Diane Symbaluk.

ISBN 0-17-625182-0

1. Social sciences—Authorship.
2. Report writing. I. Title. II. Title:
Guide to success in social science.

H62.S94 2005   808'.0663
C2005-903524-2

CONTENTS

■ Chapter 5 Referencing Using APA Format   37

■ Chapter 6 Writing Papers   43

■ Chapter 7 A Research Proposal   51

PREFACE

Thomson Nelson Guide to Success in Social Science: Writing Papers and Exams is based on my experience as a struggling undergraduate student, a determined graduate student, and a student-centred instructor. The suggestions in this guide are largely the result of best practices used by students who achieved high grades in my courses. I've also included a few guidelines that I've developed to help circumvent lessons learned through the heartache of students who tried but did not master the skills needed to achieve in one or more of these areas. It is the common errors, questions, sources of indecision, and various writing and learning styles of my students that have ultimately culminated in this book that I hope will afford all students an edge in their quest to succeed in the social sciences.

■ OUTLINE OF THE BOOK

The first chapter introduces you to this book with some general guidelines for studying and writing. Chapter 2 focuses on more specific strategies for studying, including how to translate course activities and readings into meaningful lecture notes, how to organize lecture notes into study notes, and how to study effectively. In Chapter 3, multiple-choice, short-answer, and essay questions are explained, including how the questions are developed and how best to tackle exams based on them. Chapter 4 describes methods for navigating the library and the Internet to find appropriate reference materials for course assignments. In Chapter 5 you learn about referencing using APA format, the most commonly used style in the social sciences.

Chapter 6 explains how to write an essay, including how to narrow your topic and outline your ideas. Chapter 7 describes the objectives and main components of a research proposal, including important ethical considerations in research. The main format for writing a publishable research report is outlined in Chapter 8. In Chapter 9, you learn how to start developing your curriculum vitae to convey your academic achievements to a potential employer. Finally, Appendices 1 and 2 include a sample essay and research report written by a student who earned high marks and whose writing demonstrates many of the principles in this book.

ACKNOWLEDGMENTS

I extend my first word of appreciation to my husband and life partner, Jeff Baxter, and my stepson Rob, who played in the snow, scooped with the Tonka bobcat, drove out to the lake, and otherwise amused my two-year old son Alex while I wrote this manuscript.

In preparing *Thomson Nelson Guide to Success in Social Science* (and other publications), I have benefited from suggestions, feedback, and ongoing support from a super team at Thomson Nelson including Maya Bahar, Natalie Barrington, Joanna Cotton, Daryn Dewalt, Katherine Goodes, Tara Hogue, Edward Ikeda, Sandy Matos, Nancy Mucklow, Tara Needham, Elke Price, Tammy Scherer, Erica Smith, Lenore Taylor, Tannys Williams, and Cara Yarzab.

Rebeccah Marsh also deserves special recognition as a great friend and colleague. When I first took on this project, Rebeccah helped keep me sane by partnering with me on regular runs to blow off steam and calories. Later on, Rebeccah read the manuscript, chapter by chapter, sending me e-mails that contained invaluable feedback and suggestions.

The sample essay and research report included in this guide were written by Lisa Moldaver, a former student of mine whom I hire as a teaching assistant every chance I get because she is amazingly organized and resourceful. I would describe Lisa as a very bright, practical, and grounded individual who manages her time effectively. She has graciously permitted me to use her work to illustrate an actual student's writing that demonstrates many of the suggestions provided in this book.

Finally, this book is dedicated to my mentor, Dr. W. David Pierce, a brilliant researcher and accomplished writer who taught me how to conduct thorough research and construct meaningful classroom demonstrations. Somewhere in the process, he even taught me the life lesson of how to have faith in my abilities. Over the years, Dave helped me appreciate that writing is not something you wait for (like an inspiration), nor is it a natural instinct (i.e., an inherent talent); it is a behaviour, and as such, it is something that can be shaped. And if you persist, over time it becomes something you *learn* to do well.

—Diane Symbaluk

ABOUT THE AUTHOR

Dr. Diane Symbaluk is a faculty member in the Department of Psychology and Sociology at Grant MacEwan College in Edmonton, Alberta, Canada. She is currently the Chair of the Research Ethics Appeals Board. In addition to teaching, Dr. Symbaluk serves as Resource Development Officer for the Office of the Vice-President, Resources. Among her many publications, Dr. Symbaluk has co-authored a textbook on learning and has written scholarly articles, study guides, on-line study materials, test-banks, and other ancillary materials.

CHAPTER 1

Introduction

There are no magic rules for how to achieve a high grade on an exam or how to write the perfect paper, and some of the practices you develop might be unique to your individual learning style, personality, schedule, and work ethic. Listed below are some recommendations I routinely give my students about how best to prepare for exams and writing assignments.

■ GUIDELINES FOR STUDYING

Set aside several blocks of time to review your notes and develop your study strategies well in advance of the actual exam date. You never know when you will take ill, when there will be a family emergency, or even when a work or travel opportunity will present itself and compete with your study plans. The more prepared you are at any given time, the better your odds are for achieving a high grade on the eventual exam. You can't put your life on hold, but you can schedule quite a few potential study sessions. Gollwitzer (1999) refers to these plans as "implementation intentions," and he notes that these strategies help build our resistance to distractions and help us achieve our goals. Sometimes it is a more devoted student with good intentions and good time management, rather than an exceedingly smart one, who actually gets ahead.

Learn to study in noisy places. Although it is often recommended that you choose a quiet, distraction-free study space for optimal learning, there is a reasonable chance that you will write exams in a variety of settings and circumstances. A final might be scheduled in a different room or a gymnasium. Even in your regular classroom, unpredictable distractions occur (e.g., the person next to you coughs every couple of minutes, there is an irritating sound from an overhead fan, construction noises can be heard in another part of the building). If you learn to study in different environments (noisy, quiet, with or without other people), your ability to tune out distractions and concentrate is greatly enhanced. Just ask a student who is also a single parent for confirmation of this advice.

Make notes while you read the textbook. Take down a few key points as you read the main sections of the textbook. Alternately, you may want to use a highlighter pen to underline some of the key ideas. I opt for writing out the key ideas over highlighting because I tend to get mentally lazy and highlight everything with the notion that I will eventually come back to it to decide what is important. When I take notes, I focus on the material I am reading at the time to determine the key points, and the additional activity of writing it down actually helps me memorize it. In either case, the main ideas may not be as salient to you a couple of weeks later, so take note of them while you do the initial reading of the chapter. In addition, try to paraphrase the points instead of directly copying the author's wording. This will help you to "actively study" instead of "passively study." As with exercising, simply going to the gym and hanging out with friends will not get you into better shape. If you're going to set aside the time to study, make it count.

Make sure you are aware of what is being emphasized in class. Your instructor often builds in pauses for you to take notes in class and may even repeat important points that need to be recorded. Take advantage of this time. Consider if the instructor conveyed the same information in several different ways (e.g., first she defined the concept, then she showed a video clip that illustrated that concept in Asia, then she asked for feedback about how that concept might apply in Canadian society, and so on). This is a main idea that should be highlighted for later consideration. Try to think up a way that this material could be stated in an actual exam question (e.g., what could the instructor include for a short answer question on this?). Write the question in your notes right away as it will be invaluable to you when you come back to study the material several weeks later.

Finally, carry your notes with you. A set of notes is easier to lug around than a textbook. There are all kinds of opportunities in a day to read the notes, and you can get most of your studying done during downtime from other tasks (e.g., while you wait to see your doctor, while you sit in the cafeteria and eat your lunch, and while you ride the bus home). Ideally, you want to create a shorter set of study notes that contain only the key concepts and link all the main ideas. Main points can be written on index cards for easier transport (e.g., they readily fit in a knapsack, inner coat pocket, or purse).

■ SUGGESTIONS FOR WRITING

Don't wait for an inspiration. Whether you are trying to write an essay for a class, a letter to your mother-in-law, or even a book like this one, there is no perfect time to write. It is true that you might be more energized at certain times of the day or may think more clearly if you are well rested, but keep in mind that writing is a task that takes an enormous amount of time. So regardless of how you feel about it, the more time you can put into it, the better. When you are not able to convey your ideas appropriately (i.e., you are suffering from some kind of "writer's block"), redistribute that time in a productive manner (e.g., review what you have already written, locate some appropriate references, create a title page, sketch an outline of the arguments you hope to include in the paper—all of these steps have to happen at some time before the product will ever come together anyway).

State the purpose of your writing early on. As an instructor, I do not want to have to infer or guess what you are trying to write about. If you conducted a study, you are now a quasi-expert on the subject matter, and your research report needs to detail what you've learned, as demonstrated in the following introductory statement: *This study examines the relationship between sleep deprivation and retention of detail among junior high students.* If you recently read a book on dream interpretation, the purpose of your paper should be obvious early on, as evident in this thesis statement: *This paper compares and contrasts Freudian and Jungian approaches to dream interpretation using a case description involving a man named Clark.*

Stay on task and write to the point. Where possible, state your message as clearly and succinctly as possible. Long, drawn-out attempts to pretend you are saying something are not convincing and may actually bore your reader. (Think about how you feel when an instructor gets too far off topic in class.) Great tools of the writing trade include a dictionary, a thesaurus, and an American Psychological Association (2001) Publication Manual (discussed in detail in Chapter 5).

Review the main criteria for evaluation. Based on what you have written, did you meet the objectives needed to ensure a high grade? Did you answer any questions that were provided ahead of time by the instructor? Did you clearly define the key concepts? Did you provide examples where possible? Did you check it over for spelling and grammar mistakes? Try to allow yourself sufficient time so that you can leave the completed product for a few days and then proofread it (inconsistencies and errors are more apparent once you've had some time away from your writing).

Provide all relevant information when you hand in your paper. Unless otherwise specified, your work should have a title page that includes your name, your student identification number, the date, the course you are in, and your instructor's name (see Chapter 6 for an example). Inclusion of these items is necessary for practical reasons. What if you accidentally leave your paper in another class? It is also a good idea to number the pages, staple your paper or collate it inside a duo-tang or book report cover. What if your instructor drops the whole collection of student papers?

Writing Checklist:
- Include a title page
- Use 12-point font
- Use Times New Roman (or something similar)
- Double-space
- Indent new paragraphs
- Centre first-level (main) headings
- Second-level (sub) headings are generally at the left margin in bold or italics
- Follow APA format (unless otherwise directed)
- Begin references on a new page, at the end of the paper

CHAPTER 2

Strategies for Studying

You can study for countless hours and still do poorly on an exam if your study notes are incomplete, or if you mainly study information that is peripheral to the lectures and assigned course readings. This chapter helps you create meaningful lecture notes, provides strategies for turning your lecture notes and assigned readings into study notes, and gives you tips on how to study effectively.

■ LECTURE NOTES

Read the textbook chapter or other assigned readings in advance of a corresponding lecture. This gives you a frame of reference for the lecture. A lecture is often designed to help clarify a few key issues raised in a chapter. If you are already familiar with the main ideas from the chapter, the lecture can provide you with all sorts of examples that you can later use on an exam to illustrate the key ideas or depict different points of view (e.g., as evident in comments made by the instructor and your classmates). Finally, if you read the chapter first, your instructor's own bias or focus is more readily apparent (and your exam is likely to reflect this).

During a lecture, stay focused on the main ideas that are being conveyed by the instructor. If classmates make suggestions you can use as examples of key ideas, write them down. If you hear responses that are largely off-track, try not to get distracted, and either leave the opinions out of your notes or make a reference that the comments are other ideas raised by people in the class. Then when you come back to study these notes later, you are reminded not to put a lot of emphasis on them.

How much should you write down? Probably one to three pages of notes per lecture. Write down everything the instructor emphasizes in some detail. Sometimes an instructor repeats main ideas within a lecture, writes key ideas down on the board, and shows slides using an overhead or ceiling-mounted projector. These are the most obvious sources for eventual exam questions. Any key ideas stated in this fashion should be recorded verbatim. You should also try to expand on each key idea and include a few examples where possible as these ideas might end up in short-answer or essay questions on an upcoming exam.

When you copy main points from a lecture, leave additional space for clarification. Suppose you are listening to a lecture on sexual assault and the instructor puts up a PowerPoint slide with the heading *Changes to sexual assault legislation after 1983*, with three points included under the title as follows:

1. definition
2. levels
3. treatment of victim

These are the kinds of main ideas you need to recall and elaborate upon in an exam. An expanded point you should include for the first item is the actual definition of a sexual assault (i.e., *any unwanted sexual contact*), and perhaps contrast this with the older definition so the change in legislation is apparent (e.g., *Before 1983 the term rape was used instead of sexual assault and a rape involved a man having sexual intercourse with a woman, who was not his wife, without her consent*). A few of your classmates may make comments on the old law but you do not need to include their opinions in your notes.

If a highly relevant question is raised and answered in class, include this information in your notes. Someone asks, *Under the old law, could a husband be charged with raping his wife?* Your instructor answers *no*, and points out that women were also not considered in the category of rapists. Write something like: *Note that under the old definition, or prior to 1983, a man could not be accused of raping his wife, and a woman could not sexually assault a man.* Each main point needs elaboration to prepare you for the exam. Now you can see that missing a class and borrowing the information from another student might identify the key ideas. However, without the additional explanation, you would have difficulty clarifying these points on an exam.

Great lecture notes include general reference statements that summarize special class activities and note their purpose. *What issue was highlighted in the video and how did it help explain other course material?* Suppose your instructor lectured about the main theoretical perspectives in sociology. Next, she showed a video clip on people in India who sell body parts (i.e., a kidney). Finally, the class formed groups to discuss this practice. Your notes should list and briefly outline the structural, functional, conflict, feminist, interactionist, and postmodern perspectives (i.e., the main ideas). Also, indicate that a video on selling body parts in India was shown as an example of an issue that could be viewed using any of these perspectives. Lastly, record ideas raised in the groups in relation to the perspectives. Maybe one group noted that a structural functional perspective focuses on the positive aspects for society, including extending the life of the wealthy sector (the donor recipients). Another said that the conflict perspective highlights the implications of power and social inequality since kidney donors are exploited by the rich and further disadvantaged by subsequent medical complications.

Perhaps a police officer came to your class and spoke on community perceptions of youth, the kind of crimes youth commit, and some of the methods for dealing with youth crime (e.g., Youth Criminal Justice Act, Mediation). Don't try to write down everything a speaker says. *Think about how the guest lecture fits with your course materials and try to write down points that integrate the two.* Your notes could indicate that the guest lecture expanded on the assigned readings about correlates of crime. Younger persons are more likely than older individuals to commit crime, and this is especially true for property crimes such as vandalism. The guest speaker noted that perceptions of youth are generally negative even though only a few young people are responsible for most youth crimes. He also suggested that jail terms are not particularly effective as a means of deterring crime and discussed some alternatives to prison. Your notes can contain more specific examples and details about the alternatives to prison.

Class demonstrations that sometimes resemble games are fun to play and they help illustrate important points. Suppose your instructor staged an interactive show resembling *The Newlywed Game*, where students guessed partners' responses to *What do you typically put on a piece of toast? What kind of pet is the best one to get?* If this were an introductory sociology course and I were instructing it, I would be demonstrating our *shared culture* as evident in the similar and limited responses to the questions posed. We put butter, jam, peanut butter, or cheese spread on our toast. An ideal pet is a dog, cat, bird, or rabbit. This facilitates a discussion of internalized rules called *norms* that we have in Western culture for appropriate foods to eat or pets to own. It also anticipates a discussion of *ethnocentrism*, or our tendency to judge other cultures from within our own limited perspective. Your notes should include mention of the game and how it demonstrated the concepts discussed above.

Some comments on formatting lecture notes. Write the date at the top of each page and number the pages. Point form works well and is quicker than trying to form complete sentences while the instructor is speaking. You might want to use more than one colour of pen, but try not to get too fancy. You can go over your notes later to underline or highlight the key terms. Try to develop some personal shorthand. For example, draw a star in the margin to indicate material that the instructor has stressed several times, a box around important terms, and place a question mark beside information that you do not fully understand.

■ STUDY NOTES, STUDY TIME, AND STUDY STRATEGIES

Don't focus too narrowly by trying to predict the exam questions precisely. Some instructors rely on test banks that may be authored by someone other than the individual(s) who wrote the textbook. In this case, exam questions can read somewhat different from the way your instructor presented the information in class or the way it was discussed in the text. Your best strategy for obtaining a good grade is to make sure you have clear study notes that help you understand the key ideas (i.e., don't just try to memorize them). Effective study notes also prepare you for differences in the manner in which the information is conveyed on exams.

Start with the lecture materials and try to refine them. Your lecture notes are probably fairly lengthy and detailed at this point. Read over the lecture notes you made in each section of the course and try to refine them. Rewrite your lecture notes into a shorter, more concise set that becomes the foundation of your study materials. This is also a good time to incorporate any notes or other observations you made while going over the textbook or additional readings. As you read through the notes for each lecture with your highlighter pen in hand, ask yourself the following questions:

What concept or point was emphasized in this lecture? All main ideas should be highlighted and you should rewrite these points to help you memorize, clarify, and understand the key ideas in detail. *Was a concept or point illustrated in more than one manner?* If your instructor defined a key term, then she had the class come up with examples of the concept, and then you watched a video on it, that concept is a very good candidate for a short-answer or essay question. *Did your instructor give any hints about exam questions?* Sometimes an instructor bluntly states that a certain concept or idea would make a great exam question. Or that you need to understand a particular theory and be able to apply it for your midterm exam. In this case, your instructor is not playing with your head, but rather, trying to make sure you understand that the idea is central to the course and is a main learning objective. Other times, an instructor gives more subtle hints or directives for studying by using phrases like *This is an important consideration, This is a key point,* or *This is a fundamental assumption of the theory.* At the very least, concepts, definitions, and ideas that are repeated, emphasized in more than one way, or summarized at the end of lectures should be noted for special consideration.

Pay particular attention to learning objectives. If you are lucky, your instructor might distribute a list of learning objectives—perhaps on the course outline, sometimes chapter by chapter, and possibly as study suggestions just prior to exams. Learning objectives direct you to the key concepts, people, ideas, arguments, methods, theories and points. In addition, objectives indicate how you need to demonstrate your mastery of these materials, as illustrated with this example: *Explain what C. Wright Mills meant by sociological imagination.* Here, you are expected to provide a definition of the key concept along with an example or application that demonstrates its use.

> The sociological imagination is "the ability to see the relationship between individual experiences and the larger society" (Kendall, Lothian Murray & Linden, 2004: 9). In this case, we need to consider how personal troubles relate to the wider context in which we live. For example, Emile Durkheim (1897) noted how suicide rates vary according to the degree of integration and regulation in society (quite independent of people's personal issues).

This learning objective could directly appear on an exam as a short-answer question likely worth about three marks, or it could be adapted into one of these multiple-choice questions:

1. _____ coined the term *sociological imagination*, or the ability to see an inter-relationship between personal issues and the wider society.
   a. C. Wright Mills*
   b. Emile Durkheim
   c. Talcott Parsons
   d. Auguste Comte
   e. Herbert Spencer

2. The ability to see an interrelationship between personal issues and the wider society is known as
   a. the sociological imagination.*
   b. positivism.
   c. social solidarity.
   d. symbolic interactionism.
   e. the socioweb.

3. Viewing suicide as a societal problem that occurs when social regulation becomes too excessive demonstrates
   a. the sociological imagination.*
   b. positivism.
   c. social solidarity.
   d. symbolic interactionism.
   e. the socioweb.

If there is an accompanying study guide, use it. A study guide is a great learning tool for students who benefit from instructions on how to structure study notes and from examples of how course materials translate into actual exam questions. Most study guides include an outline, a summary, a list of key terms and people, some learning objectives, and a set of quizzes that correspond to each chapter of the assigned textbook. The outline sketches out the chapter contents while the summary alerts you to key issues and provides you with a framework for organizing the contents. The list of key terms identifies the main concepts. Key people are the main theorists whose ideas are central to that chapter. Learning objectives identify what you are expected to get out of your assigned readings and should form the foundation of your study notes.

Finally, quizzes test your understanding of the learning objectives and may be included in a variety of forms. Multiple-choice questions test your ability to recall facts, define key concepts, and apply the concepts in a meaningful way. True-or-false questions determine if you know about and understand key facts. Fill-in-the-blank statements are generally used to identify key concepts, facts, or people. Matching problems help you connect theorists to their main ideas, and concepts with their appropriate theoretical perspectives or originators. Short-answer and essay questions invite you to explain key concepts and demonstrate your understanding of the learning objectives in more detail.

Taking these little practice tests can help you to determine any subject areas in which you are weak, as well as the type of questions with which you might have difficulty. For example, you might need to reread your notes from Chapter 4 or practice choosing quick and correct answers to multiple-choice questions.

How much time should you spend studying? Only *you* can determine how much time you have to allocate and how much time it takes you to read, write study notes, and learn the material. Irrespective of the actual amount of time, use the format of the exam and any hints or learning objectives provided by your instructor to help you distribute your study time most effectively. Suppose your instructor notes that the exam has 60 multiple-choice questions that correspond to five chapters of readings. You can translate that information into about 12 exam questions per chapter (a few chapters might have been given more consideration and you can factor that in). You can also reasonably expect a couple of exam questions on each main idea discussed in class. If your instructor focused on about four main issues, theories, or sections per chapter, you have accounted for 40 of the 60 questions.

The vast majority of your study time should be spent studying the guaranteed items. When you know this material very well, you can use any remaining time to review text materials your instructor did not cover in class or other points that might constitute the four remaining questions per chapter.

Should you meet with your instructor for help before an exam? Wait until you have prepared your study notes and have practiced the material, and then go see your instructor if you feel additional help is warranted. Otherwise, there is no context for the visit, and it is a waste of time for both of you. A lack of preparation results in vague questions such as *How much should I study from Chapter 3?* and *Do I need to read everything in Chapter 1?* If you made detailed study notes based on the learning objectives, you'd already appreciate that only a few questions might come out of the extra 20 pages in Chapters 1 and 3. Thus, the question of whether you should read everything can only be answered by *you*, given the most appropriate use of your time.

Demonstrate your understanding and then ask how the material might translate into exam questions. Perhaps in the process of creating study notes, you encountered one learning objective that wasn't particularly clear, and you want to seek clarification about that item. First, demonstrate your understanding of the issue and then ask for more information on it. *Here's what I understand about C and D. Can you clarify the main distinction between C and D in case I encounter an exam question based on it? I'm having trouble distinguishing between these two concepts—can you provide me with some additional examples that would help me answer an exam question on this?* You might even try making up a sample exam question. *I've made up a question for this section. Can you tell me if this is the sort of question that I could expect to find in the short-answer section of one of your exams?*

Ask your instructor to take a quick look at your study notes. Your instructor might be willing to provide you with some feedback about whether your study materials appear complete, given the learning objectives. In this case, you are actually determining if you would obtain full marks if you wrote answers based on the information you have in your study notes. Your instructor may direct you to areas that need more clarification prior to the exam and this provides additional hints about potential exam questions, especially ones that are less central to the main ideas.

What are some other study tips you can try? The study strategies you select and the sources you find most effective likely reflect the kind of learner you are (e.g., visual) and the amount of time you can commit to your studies. Here are some study methods you can try out:

Use Web resources. Sometimes the textbook publisher provides you with free access to an accompanying website that may have student resources (similar to a study guide) including quizzes or summary information on each chapter. The more ways you encounter the same material, the greater your odds for understanding and repeating it on exams.

Make up index cards. Since learning objectives are central, you can write out the answers to learning objectives on index cards. Index cards readily accommodate definitions and key assumptions of theories, and they can be used to match people to the main ideas that they originated or developed. Index cards are great for testing your recall ability (since you can write key terms on one side and their corresponding definitions on the back). Index cards travel well and can be stored in your purse, in your backpack or even in your pocket for easy reference at times when it becomes convenient to study (e.g., you are riding the bus home from school, you are waiting to see a physician).

Create outlines and diagrams. Sometimes students get caught up in the details and lose track of the more general framework for the readings. One way to keep the main concepts, theorists, and ideas in order is to create an outline that provides a brief overview. This is important, as many answers to multiple-choice questions are based on the understanding of main concepts and ideas. In addition, short-answer and essay responses flow better if you study from an outline that includes the key points for elaboration. Create an outline for each chapter of the textbook. You can copy the one provided at the beginning of most chapters and then revise it to include other concepts or ideas introduced by your instructor in lectures. You might also find it helpful to draw a flowchart or diagram to clarify the relationships between key points or concepts.

Invent retention and retrieval strategies. Can you make up a silly saying, or develop some sort of personalized phrase that corresponds to the first initial of each word in a stage or component of a theory? For example, in trying to remember Hirschi's (1969) social bond theory, you might remember a saying like *Al can be involved* to remind you that the four bonds (attachment, commitment, belief, involvement) start with a, c, b, and i. One student said she used *BICA* because it was close to the Spanish word *beca* (meaning scholarship), with which she was already familiar. As another example, to distinguish between prescriptive norms (that tell you *what to do*) and proscriptive norms (that tell you *what not to do*), you might link prescriptive to prescribe (as in medicine, where you take two pills as something you *should do*), and proscriptive to prohibit (meaning something you *should not do*).

Recite your study notes aloud. This might drive your roommate crazy (and exclude you from studying in the library), but it also helps transfer facts into long-term memory. Actors repeat their lines in order to learn their roles. After reading a section of your notes aloud a few times, glance at a heading and then try to talk about the information that follows without looking at it. If you are unable to say anything without looking at your notes or index cards, you know you are not adequately prepared for the exam. Simply reading over the materials does not guarantee that you can reproduce that information in the absence of notes.

Find study-partners. You can compare study notes, discuss hints provided by your instructor, and talk about the main issues raised in videos or other special class activities. Your group might test each other using index cards or the study notes developed around main concepts and ideas. Finally, you can exchange potential exam questions with your study friends (e.g., everyone could make up five questions per chapter). Try to coordinate study time with other regular activities, such as lunch and coffee breaks or exercise outings (you can discuss a lot of material during a run). Consider it your social time, and you can greatly improve your grade while you have fun.

Talk about what you are learning in class. A mom never grows tired of hearing about what you learned in class. Your spouse may even feign interest. One student said that talking about concepts learned in class around the dinner table helped to affirm them in her mind.

Refer to other resources to help you develop study skills. Most libraries have books on how to improve your note-taking and study habits. For a guide on how to develop other kinds of skills, including reflection, effective reading, and critical thinking, look for Collins and Kneale's (2001) *Study Skills for Psychology Students: A Practical Guide.*

Writing Examinations

Exams and course requirements tend to vary considerably by course type, course level, and individual instructor. An introductory psychology class might require students to write three exams of similar format that are a mix of true-and-false, multiple-choice and short-answer questions, while a third-year theory course in psychology might entail writing essay-based exams and a lengthy research report. Notwithstanding, an instructor's own preferences might override what is typical for the level or type of course—some instructors prefer to assign papers to ensure a written course component while others opt for more objective test measures, such as multiple-choice and true-false questions. Multiple-choice, short-answer, and essay exams are all described in detail in this chapter, along with tips for achieving your potential in each format.

■ SPECIAL CONSIDERATIONS

In some cases, students have physical disabilities or other challenges that need to be taken into consideration in the context of writing exams. For example, someone with sight limitations may need to have exam materials transcribed onto audio cassettes and may need to write the exam in a different location in order to accommodate the use of audio equipment (e.g., a student resource centre). As a student, it is generally your responsibility to inform instructors as soon as possible (i.e., at the start of the course) of any special needs or requirements so that appropriate arrangements can be made well in advance of exam dates.

Even if you have some kind of temporary impairment that might disadvantage you relative to your peers in an exam situation (e.g., a wrist or hand injury), you should contact your instructors as soon as possible to determine if an alternative is available (e.g., services for students with disabilities). Sometimes students are permitted to write exams under supervision in a resource room where extra time is allotted to make up for an injury, or a scribe may be available to write the answers down for the student with a hand injury.

■ MULTIPLE-CHOICE EXAMS

You are highly likely to encounter multiple-choice exams. As much as I hated writing multiple-choice exams in my undergraduate days, as an instructor, I often rely on them as a testing tool for very practical reasons. Multiple-choice exam questions take a long time to create, but they can be used to assess student competency across a diverse range of course content and learning objectives (Nitko, 1983). Multiple-choice items can also be completed fairly quickly by students, and this is especially relevant if a class is only 50 minutes in length. If properly constructed, multiple-choice exams provide a more objective measure of students' overall understanding of course materials than can be accomplished using essays (Cheser-Jacobs & Chase, 1992).

■ TYPES OF MULTIPLE-CHOICE QUESTIONS

Multiple-choice questions usually test cognitive ability in one of these areas: knowledge, comprehension, and application. Knowledge-based questions prompt a student to state the name of a theorist, recall a fact, or define a concept from lectures and assigned readings, as illustrated below:

1. _____ is credited with having coined the term *sociology* and is considered to be the founder of sociology.
   a. Max Weber
   b. Herbert Spencer
   c. Emile Durkheim
   d. August Comte*
   e. C. Wright Mills

2. According to the Census, slightly more than _____ of Canadians are bilingual.
   a. 58%
   b. 42%
   c. 36%
   d. 17%*
   e. 5%

3. _____ are collective ideas about what is right or wrong, good or bad, and desirable or undesirable in a particular culture.
   a. Beliefs
   b. Norms
   c. Values*
   d. Assumptions
   e. Expectations

Comprehension questions test the ability to interpret and recall key ideas and are often found in multiple-choice questions that require some kind of summary or explanation.

4. Which of the following statements accurately summarizes Robert Merton's (1968) strain theory?
   a. Repeated violations of cultural norms leads to labeling and the eventual development of a permanent deviant identity.
   b. The discrepancy between emphasized cultural goals and the legitimate means for obtaining them can lead to less conventional modes of adaptation, such as stealing.*
   c. The positions in a society that are most important require scarce talent or extensive training and thus should be the most highly rewarded.
   d. Deviance is functional for society because it can promote boundary clarification and social unity.
   e. People tend toward crime and deviance when they are relatively free of conventional bonds to society in the form of attachment, commitment, involvement, and belief.

Finally, application-based questions involve demonstrating how a concept, perspective, or theory works in a real-life context. Sometimes a question presents a scenario requiring an inference of the correct concept, or the question might ask for the best example of a given concept. An application question may even test critical thinking through the analysis of relationships or connections between events and ideas, as shown below.

5. Canadians may believe in the practice of *monogamy*, but many of them cheat on their spouses. This demonstrates a discrepancy between
   a. ideal and real culture.*
   b. material and nonmaterial culture.
   c. folkways and mores.
   d. formal versus informal norms.
   e. popular versus high culture.

Chapter 3: Writing Examinations

6. Choose the best example of a *prescriptive norm*:
   a. talking to your friend while your instructor is lecturing.
   b. looking at a friend's paper during an exam.
   c. putting up your hand to ask a question in class.*
   d. not reading a magazine during class time.
   e. giving a friend a dirty look for talking to you in class.

7. Emile Durkheim is to _____, as Karl Marx is to _____.
   a. social facts; alienation*
   b. natural selection; social disorganization
   c. means of production; social consensus
   d. value-free; meritocracy
   e. power; equality

■ TIPS FOR WRITING MULTIPLE-CHOICE EXAMS

Read the question and try to answer it before you go through the responses.
If you know the correct answer, it should stand out. If you are permitted to
write in the exam question booklet, circle the correct response when you
get to it. When selecting from among the responses, you should also consider why a given response is correct as well as the rationale for why alternatives are not (as you determine that a given response is incorrect, you can
mark it with an X or cross it out). This practice saves you time later on,
since you will only have to choose between a couple responses if you have
to come back to that question. Transfer answers from the question booklet
to the answer sheet as you go along. That way you can more accurately determine how much time you are spending on the questions (and you have the
answers already recorded if you do fall short of time).

Distribute your exam-writing time appropriately. Wear a watch to the
exam. For a multiple-choice exam you will be allotted about one minute
per question. Complete ten questions and then check to see how long it took.
If more than ten minutes passed, speed up, mark any questions that you have
trouble with and come back to them later. The first time you read a question, put an answer down for it. You never want to leave a blank on a
multiple-choice test in case you run out of time and can't go back to fill it
in—in this case, a guess made while pondering the question is better than
nothing. Also, forgetting to leave a blank can cause you to mix up subsequent items (i.e., if you put the next answer in place of the blank).

Don't change a bunch of answers. I have monitored this practice for
years, and based on outcomes and comments made by students who change
answers, the general pattern is something like this: If you change one answer
because you went back and reread a question more carefully and realized
you missed something, changing that one item generally works to your

advantage. However, if you change two or more questions, it is likely because you are not well prepared for the exam, or you are nervous, and in going back over the questions, you started to read too much into them. In either case, changing more than one answer usually lowers your overall grade. Most multiple-choice response sets include the correct answer and one that is reasonable but not the best answer. If you over-analyze the questions, you might convince yourself that the less central response is really the best one.

Don't look for patterns. I thought there should be more "c's." There were too many "a's." A "b" was due. When I make up an exam, I start with more questions than I need and I eliminate items until I get the desired length. I never look to see if there is a response pattern (i.e., I could have inadvertently eliminated all the questions that had a "b" as the correct response). Moreover, I move around bunches of questions to create multiple versions of the same exam. Questions are grouped on the basis of how they fit together (e.g., questions that relate to Chapter 2, a set dealing with a video). I never look to see what the correct responses are until I have already sent the exams for printing and am making up my answer keys for marking purposes. The exam is already in its final form when I discover that there are in fact a lot of "a's" in a row and few or no "b's."

Guess if you have to. If you can narrow it down to a choice between two answers, consider that the correct answer tends to be the longer choice, a grammatically correct statement, not an extreme value, and not an exception. If the question asks: *What percentage of people agreed that prostitution was a serious problem in society?* A good guess would be 63% out of a choice of 2%, 30%, 50%, 63%, or 100%. Similarly, if a question states, *An _____ variable is the cause in an experiment*, you should choose independent, over dependent, moderator, or predictor as your response, given that it is the only grammatically correct response. Finally, if despite everything I've said above you are still compelled to take a guess based on response patterns, Mentzer's (1982) analysis of multiple-choice exam test bank files indicates that "c" tends to be over-utilized, as does a response that contains the phrase "all of the above."

Be careful not to mark up the answer key. Most multiple-choice tests are computer-scored. If you put pencil marks on the answer sheet (e.g., you circle items that you want to come back to and forget to erase these completely), the computer detects "extra lead," assumes you gave more than one response, and marks the question wrong! If permitted, put all of your marks or comments on the actual question booklet. If you are not allowed to write on the question booklet, write the number for any items with which you had trouble on the back of your short-answer sheet or on a scrap sheet of paper (if you receive one for sketching out ideas). You could even ask in advance of the test if the instructor can bring some blank sheets of paper to distribute so that students can write down extra thoughts, outlines, or question numbers during the exam.

◼ SHORT-ANSWER QUESTIONS

Short-answer questions are common on social science exams because they necessitate some writing on the part of students, and they are perceived to involve higher order critical thinking that may not be demonstrated via multiple-choice and other objective tests. In most cases, short-answer questions prompt students to recall key ideas and theorists, describe main issues, present information using a theoretical perspective, or express ideas in their own words. Since only a few central ideas or concepts can be measured, and scoring is much more subjective, an instructor may opt for a combination of both multiple-choice and short-answer items on the exam.

If your test combines multiple-choice and short-answers, you can take a moment to quickly skim over the short-answer questions (and even jot down a few ideas), but *start with the multiple-choice section*. It could prove disastrous to your grade if you start with the written section, lose track of time, and have to forfeit marks in the multiple-choice section. Remember, multiple-choice items correspond to about a minute per question (so if you have a 90 minute class and 30 multiple-choice questions, you know you should spend no more than 30 minutes in this section, leaving you an hour for the short-answer portion). Sometimes the practice of reading and answering all of the multiple-choice items even helps you identify or recall information you want to include in your short answers.

◼ TIPS FOR WRITING SHORT-ANSWER EXAMS

When you get to the short-answer section, *jot down in point form some relevant concepts, people, components, steps, or stages for each question*. Mini-outlines help you plan what you need to include and help keep you on track with your responses. Put them on the last page of the exam booklet so your instructor sees only the carefully constructed, full responses. If you think you might fill an entire booklet, ask for a spare one and write your notes in the spare booklet. This practice is important, because if you run short of time, you can number your outlines and turn them in for partial marks. If you write throughout the exam period and still run out of time, it is likely that others have met the same fate, and your outlines can be used to demonstrate that you understood the material.

Start with the short-answer question you are most comfortable with. If you run out of time during an exam, you want to get full marks for everything you wrote up until that time. Save the hardest question for last, since you may need time to ponder the answer, and if the answer never comes to you, you already have the most marks you could possibly obtain up to that point. Also, starting with the easiest question will help boost your confidence and get you into speed-writing mode.

Read the directions carefully because the wording indicates what you need to accomplish with your answer. For example, the words "list" or "identify" instruct you to put down the key points. *List the five primary taste sensations. Identify the beliefs that are shared among cult members who commit mass suicide.* The term "describe" suggests that you need to explain something in more detail, often with the use of examples. *Describe the two types of brain degeneration that occur with Alzheimer's disease.* "Define" is a common question root that involves explaining the meaning of something. *Define infantile amnesia.* "Summarize" prompts you to write all of the main ideas or assumptions, while "discuss" suggests you should write the key points and some additional information (such as examples, support for this view, etc.). *Summarize the conclusions of Rosenhan's study. Discuss the link between mental illness and homelessness.* "Compare and contrast" generally means you should state all of the similarities and differences between two things. *Compare and contrast Battered-Wife Syndrome and Self-Defence.*

The total number of marks for a question can be used to determine the length of your response. A question that is worth one mark might prompt you to write a single term or a phrase without elaboration. *What is the correct term for a mental shortcut that involves judging frequency on the basis of how salient an event is in your memory? What type of non-probability sampling procedure was used to obtain participants for an exploratory study on prostitution?* A question worth two or three marks generally requires a two or three sentence response. *List and describe a direct observational method for obtaining information on crime. Identify the potential responses to being stigmatized.* A question worth a total of five marks is likely to generate an answer that is at least a couple of paragraphs in length and may be fully explained in a few pages of writing. *Compare and contrast Battered-Wife Syndrome with Self-Defence. Describe each of the five modes of adaptation included in Merton's strain theory and give an example of each.*

■ ESSAY QUESTIONS

The highest level of critical thinking, according to Bloom's (1956) *Taxonomy of Educational Objectives*, incorporates analysis or evaluation without a direct link to course materials. Consider the following essay question for a final exam in a deviance course: *Apply the three components of the social typing process to one substantive issue that has been addressed in the textbook.* Essays are logically constructed positions that discuss key theories, apply key concepts, and summarize main issues with the aid of supporting examples. Essay questions necessitate considerably longer responses than short-answers. Essay questions may be given as take-home, open-book, or regular in-class examinations.

A take-home essay is similar to a research paper. Students erroneously believe that a take-home essay is easy because you have the questions ahead of time. Take-home essay exams require a great deal of synthesis and analysis and may necessitate more preparation than is required for an in-class exam. Problem-solving skills are also being tested, since take-home answers usually involve the integration of relevant information from lecture notes, text chapters, and other reference materials similar to those needed to write a research paper. Use the advice provided in Chapter 6 to help you write a take-home essay.

Don't underestimate an open-book essay exam. Open-book exams enable students to utilize any or all of their course materials while completing the exam. An open-book essay exam appears to be fairly simple in the sense that all of the needed materials to answer the questions are readily available, and there is no need to extract details from memory. However, open-book exams require a great deal of writing and there is not adequate time to relearn information or look up every relevant detail at the time of writing.

To prepare for an open-book essay, write out concise, easy-to-read notes ahead of time. You need to rely heavily on your notes while you write an open-book exam, and that means the notes you make ahead of time strongly correlate with the grade you achieve on the test. Write out definitions for all the key concepts and create an outline of key points for each chapter. *Devise a method for organizing your notes so you can refer to a needed section quickly* (e.g., use tabs, highlighting, index cards, some kind of number scheme). The remaining sections discuss strategies designed to help you write the more common in-class essay exams.

◼ TIPS FOR WRITING ESSAY EXAMS

Use a pen, double-space your answers, and write as legibly as possible. The astuteness of an extremely difficult-to-read essay may be unintentionally underrated by an instructor. Consider that it may take several hours and even many days for an instructor to mark the essays from a recent exam. An instructor grows weary of trying to find insight among pencils smudges and other illegible forms of writing, so the more clearly you can present your ideas, the better. First-rate presentation leaves a favourable impression on your instructor, and it makes awarding marks for appropriate content much easier.

Choose the best essay question from a set of questions. Sometimes you can choose to answer one essay question from among a set of questions. *Please select and answer one of the following four essay questions.* If you are instructed to select one of four choices, be sure you only write on the one you have the most knowledge about. If I ask students to select one choice and more than one is actually answered, I only mark the first one (and cross out the other responses), even if it is apparent that a later choice is answered more thoroughly or correctly.

Make sure you understand what is required in the answer. Sometimes you are prompted to supply several things in your answer. If you are asked to "describe" key concepts and "provide examples," you will achieve only partial marks if your essay describes the concepts but fails to include any relevant examples. If you are prompted to "analyze," "explain," or "discuss" something, it generally means you need to tell the reader about an issue or event in detail, noting the key concepts, theoretical perspectives, and main assumptions. Elaborate on what you would have included for a short answer and assume that you should define all relevant key concepts where possible.

You might need to render a judgment. In some cases, you might be asked to render a judgment in your answer, as in this question from a criminology course: *Does Hirschi's social bond theory explain vandalism?* You need to include a lot more than just a direct response to the question posed. *Yes, Travis Hirschi's (1969) social bond theory can explain vandalism of bus shelters because youth who ruin bus shelters tend to be relatively free of the conventional bonds that inhibit others from committing similar acts.*

In addition, you want to make sure you note the key assumptions of the theory (i.e., *Social control theory assumes that all people are motivated to commit crime. Social bonds help people to refrain from committing crimes*). Also, where possible, include definitions of key theoretical concepts (i.e., *bond elements of attachment, commitment, involvement, and belief*). Try to provide some relevant details (i.e., *information on vandalism*). Finally, organize the parts of the essay in a logical fashion. For the essay question posed above, you might conclude by summarizing your overall evaluation of the adequacy of control theory for explaining vandalism. Some of the necessary components may be prompted by instructions on the exam: *Be sure to describe the key assumptions of the theory, define the main concepts, and provide examples where possible.*

Create a mini-outline of what you plan to include in your essay. Quickly write down any key ideas in point form. Note any key theories and associated theorists you plan to incorporate in your answer. List the key assumptions of the theories or write down any key concepts that need to be discussed in some detail. This process helps you identify main components you need to work into your essay regardless of how you answer the question posed. The mini-outline also ensures you a few marks in case you run out of time (as noted in the section on short-answer exams). Write your essay using the mini-outline as a guide.

Begin with a strong opening statement that addresses the essay question. For example, you could answer the question, *Does Turk's theory predict what happened during the L.A. riots?* with *Austin Turk's early conflict theory strongly predicts clashes that occurred between police and rioters in Los Angeles in 1994.* Sometimes there is no absolute, correct answer, and you are likely to earn most of your marks for including relevant concepts and for your explanation, so take a stand and write to strengthen that position. Also, write to the point. Your instructor does not want to hunt for the main ideas or try to infer your position. Demonstrate your understanding of all of the key concepts by defining them and providing examples.

Finally, *write something down even if you can't answer the question precisely* (i.e., you really don't know what the correct answer is). It might turn out that a question was a little vague or the majority of the class was unable to recall the main idea or concept the instructor was seeking in the answer. As long as you made an attempt, you are likely to receive some credit for it (i.e., an instructor who is trying to be lenient needs something to apply marks to). If you leave an item blank, you have forfeited any grace an instructor may be able to afford you. Consider how important this might be if *one* extra mark on this exam would have changed your overall percentage and ultimately earned you an A instead of an A- as a final grade in the course.

CHAPTER 4

Finding Information in the Library and on the Internet

Can you get all the reference materials you need for your assignment over the Internet? No, you cannot. You still need to visit an academic library. Libraries are superior to the Internet for a number of reasons; two of the most important ones involve quality control and copyright restriction (Mann, 1998). Anyone and anything can be published on the Internet. Some of the information you find on the Internet may be inaccurate, uninformed, and misleading. A library holds a large variety of resources, including books, journals, newspapers, magazines, government documents, art slides, video recordings, DVDs, music, music scores, computer software, and electronic resources. Library staff select resources based on reliability, currency, and relevance to an institution's curriculum. The resources are permanent, organized, and free for use by patrons. They also come with personal assistance from the reference staff. Finally, aside from storing enormous holdings of reference materials, libraries provide quiet study areas, photocopy services, and in some cases, technology resources such as access to computers, video equipment and database facilities.

The library you are using is likely divided into a number of areas, including reference, technical services, and circulation. The reference area is the central place for library instruction on how to locate appropriate materials for class assignments. Trained experts help library users search topics, refine topics, and find information using the online library catalogue, reference materials, and electronic databases. College libraries usually have some kind of technical services that can assist you in resource bookings, such as signing up to use a computer for Internet use or word processing. Printing and photocopying services may be handled by this section as well. Visit the circulation area before you exit the library in order to secure borrower privileges, sign out books and other library materials, and obtain reserve materials.

Where possible, try to make use of scholarly articles and books as your reference sources. The exact nature of materials needed for an assignment depends on the scope and size of the project. A 750-word (i.e., 3-page) essay describing "Milgram's Classic Study on Obedience to Authority" may only require Milgram's (1963) article depicting his first experiment. Conversely, if the assignment is a 10-page essay describing "Obedience to Authority," you might define various forms of compliance, note the roles of informational and normative social influence, describe Milgram's various experiments, and provide some real-life examples. In this case, your references could include Aronson, Wilson, Akert, and Fehr's (2004) textbook on social psychology, a few of Milgram's (1963; 1976) journal articles, Milgram's (1974) book on obedience, other articles written about Milgram's studies (e.g., Blass, 1996; Miller, 1986), and more contemporary articles on obedience (e.g., Meuss & Raajmakers, 1995; Moscovici, 1985). The rest of this chapter helps you locate the best sources for your course assignments.

■ SEARCHING FOR BOOKS

Most libraries have online catalogues that you can either search using a computer in the library, your home computer, or a computer, in a lab (as long as you have a student card that allows you access). An online catalogue is usually set up to search for holdings by author, title, keyword, subject, or some other kind of specialized search (e.g., series, call number, song title, ISBN). Use the menu to search as indicated below.

Author—To begin a search by author, type in the full name of the person who wrote the book (by last name first). With this search, you can obtain the list of works by a particular person as contained in that online catalogue (indicating the books by that author available in the library you are using). You could try "Milgram, Stanley" to see if the library contains a copy of Milgram's (1974) book for an essay on obedience. If you do not know the complete name of an author, most online catalogues allow you to conduct a broader search using "Author—keyword," where you type in only the author's first or last name. If you type "Milgram," you would likely obtain the same search results as with "Milgram, Stanley" (i.e., only a short list of authors has that last name). However, if you can only recall the name "Stanley," you could end up with several hundred matching titles.

Title—To conduct a search by title, type in the exact name of the book you wish to locate. This search results in a list of all of the works with that title in that library, along with the author, publisher, and call number. In this case you would type: *Obedience to Authority: An Experimental View.*

Keyword—To search the catalogue by keyword, type in a key or main identifying word in order to generate a list of books. Because the system searches for the key word in fields including title, author, contents, summary, or subject, it often results in a very long and sometimes uninformative search list. A search using "obedience" could result in several hundred titles, some that have nothing to do with your intended essay topic (e.g., I located a book on virtual art that had a chapter including the phrase "obedience to presence," and a religious book with a chapter content statement about "obedience to God"). If you end up with too many matching results, try a subject or advanced search as described below.

Subject—Using the subject index, type in a subject term to obtain a list of all titles that contain that subject as a main descriptor for the work. A search using "obedience" is not likely to match to Milgram's (1974) book, because the main subject matter is linked more to authority and experiments than to obedience (i.e., *Obedience to Authority: An Experimental View*). A subject search is great for finding out about general topics because this index classifies topics into subgroups. For example, if you wanted to write an essay for an English class on something to do with dogs, you could type in "dogs" as a subject and find holdings in areas related to dogs (e.g., *dogs portrayed in juvenile fiction, dogs as portrayed in legends*).

Advanced search—Some databases allow you to refine search interests to narrow the possibilities within an area by combining author, title, and subject keywords. In this case, you could type in "Obedience" as a title term along with "Milgram" as an author term. This search may turn up Milgram's (1969) video recording on the obedience experiments he conducted in the early 1960s. In addition to the search terms, you can sometimes refine searches according to other requirements, such as the format of the holdings (e.g., you might only want video recordings, not journals or books).

Throughout your search process, write down the call number of any books that are of further interest to you. See below for a discussion on call numbers and how they provide the address you need to locate books within the library.

■ FINDING BOOKS ON THE SHELF

Most universities and colleges shelve their book holdings using the Library of Congress Classification System (LC Classification). The LC Classification system was named after the Library of Congress, where the system was developed in the late 1890s to accommodate a move to a new building that required the reorganization of close to one million holdings (Chan, 1999). The Library of Congress is located in Washington, DC, and it is the largest library in the world with a current collection of about 128 million holdings. (You can visit the home page at http://www.loc.gov.) For a very detailed explanation of how the LC Classification system originated, how it has changed and how it arranges items, refer to Lois Mai Chan's (1999) *A Guide to the Library of Congress Classification*.

The LC Classification system groups together information of a similar topic by combining letters and numbers in what is referred to as a "call number" (Reed & Baxter, 2003). Each call number consists of a capital letter, whole number, decimal extension, item number, and publication date. The call number begins with a letter that denotes a general subject category. For example, class *A* refers to *General Works*; *B* is *Philosophy, Psychology, and Religion*; *H* is *Social Sciences*; *R* denotes *Medicine*; and so on. This means that you can walk into most libraries, and head for the *H* section to explore the *social science* holdings.

Within the social sciences, there are subclasses for main topics such as *HM* that includes holdings for *general works* and *theory of sociology*, or *HN* that has books on *social history*, *social problems*, and *social reform*. Finally, within a subdivision, you can find an even more specific topic. For example, you might find books on *drug abuse, crimes, police*, and *punishment* within the *HV* section that deals more generally with *social pathology, public welfare*, and *criminology*.

The rest of the call number helps to further distinguish items so that very specific topics are shelved together, and it helps to order individual items for easy retrieval. To find a guide on library research with a corresponding call number of Z710 .M23 1998, you first go the stacks where Z is located (they are in alphabetical order), then narrow your search to shelves containing a range of numbers including 710 (they are in numerical order), stop at Z710, now look for the "M's," and then continue until you get to Z710 .M23. Select Mann's (1998) *Oxford Guide to Library Research* off the shelf and proceed to the circulation area to sign it out. Moreover, once you locate that book, the same shelf will contain similar items that might also be pertinent to your search, including, in this case, Horowitz's (1984) *Knowing Where to Look: The Ultimate Guide to Research*.

If you take a book off the shelf but decide you do not want to sign it out, please let the library staff reshelve it. This is not an inconvenience, and it ensures that all patrons will find books in their respective locations, as indicated by call numbers.

■ SEARCHING FOR ARTICLES

Most academic libraries have some licensed electronic databases (called electronic indexes) that can be accessed within the library to help you locate scholarly or popular articles. Libraries generally pay annual fees for patrons to gain access to indexes over the Internet, or they pay for copies of print indexes that become part of the individual library's holdings. Usually, a library has a number of databases, and they will be listed on the library's homepage. Databases may provide access to entire articles (called full-text) (e.g., Social Science Full-Text, Early Canadiana Online, Statutes of Canada), partial full-text (e.g., Expanded Academic ASAP, Social Sciences Index), citations and abstracts (e.g., Medline), abstracts only (e.g., Criminal Justice, Sociological Abstracts, Social Science Abstracts), or some combination thereof (e.g., Education Resource Information Centre). The process of searching an electronic index is described in more detail below to give you a sense of how to use a database and how databases are important to your search for articles.

A full-text database includes complete articles from journals that span the area of coverage (i.e., the social sciences). Most databases include some kind of search software (e.g., ProQuest Searchware) that makes finding articles fairly easy. Simply start up the search software and then type in a search topic and press ENTER. For example, if you are interested in applications of Gottfredson and Hirschi's (1990) general theory of crime, you can type in "general theory of crime" to obtain a list of relevant articles and book reviews beginning with the most current. The list includes the title of each article, along with the journal, the year of publication, and the page number. The search software may include an icon that allows you to limit your search to only peer-reviewed (scholarly) articles.

Suppose you are interested in *Organizational offending and neoclassical criminology: challenging the reach of a general theory of crime. Criminology, Aug. 1996, p. 357–382.* The search tool likely includes an icon (such as the word "text") that you can click on with the computer mouse to see the full article or citation. Sometimes the library has the database on CD-ROM. In this case, the search tool will include a feature such as another icon that directs you to the CD on which the actual article is located. For example, there may be 30 corresponding CDs for the year 1996, and the example might be located on disk 20. After you select disk 20, all you need to do is put it into an image drive and the full-text article will appear on the screen for you to read or print.

Perhaps you have a really great article that sets the foundation for your essay or research report. The best option for identifying additional relevant articles on this topic is to determine whether other researchers have already expanded on this original work. An electronic citation index search begins with a known source, such as a journal article, and it provides you with later works that have cited the original source in a discussion, footnote, or list of references (Mann, 1998). In the absence of a citation index, take a look at the reference sections in the articles and books you have so far. Sometimes, the great sources for other related materials on your essay topic can be found in these sections.

■ THE PROCESS FOR OBTAINING ARTICLES

If the journal is available in full-text online in a database, you can generally print, download, or e-mail the article to yourself. The library usually charges a fee per page printed from a database, but you may be able to download the article onto your own CD or e-mail it to your home or school internet address so you can read it or print it off at a later time.

If the article is not available in full-text, you need to locate the actual journal within the library. Once identified, use the library's online catalogue to title search the name of the journal that has the article in which you are interested. This search indicates whether the journal is contained in the periodicals collection within the library you are using and whether the issue is available. (Usually the search provides the volume and years of the periodical contained in the library.)

If the journal is in print form, the library shelves the periodical that contains the article. In this case, you need to locate the bound or loose journal from an area containing periodicals listed alphabetically by journal title or by call number. For example, to find the article *Organizational offending and neoclassical criminology: challenging the reach of a general theory of crime*, you need to first locate the periodical *Criminology*, find the relevant volumes for *1996* and then look for *August*, and turn to pages *357–382*. You can read or photocopy the article in the library, but you cannot sign out the journal.

Older periodicals are sometimes transferred to microform to save storage space. (Search results indicate if a periodical is in this format.) If the article is on microform (saved as either microfiche or microfilm), you need to retrieve the appropriate film inter-filed in the periodical collection and display the article on a viewing screen using a microform-reader/printer. You can then browse the article (and sometimes print it) within the library. Ask for help at the reference desk if the article is on microform.

■ WHAT IF THE LIBRARY DOESN'T HAVE THE MATERIALS YOU NEED?

Inquire about inter-library loans at the circulation area. If the library you are accessing does not have an article or book you need, you might be able to get it through an inter-library loan (ILL). Many libraries participate in this inter-library lending program that allows libraries to share information among patrons (Horowitz, 1984). In some cases, there is no handling fee, and you can receive a copy of an entire journal article that you can keep for free. Keep in mind that this process can take a few days, yet another reason to start working on a paper as soon as it is assigned.

■ OTHER LIBRARY TIPS

Consider taking a library orientation session. A general library orientation session may acquaint you with the main areas of the library, help you choose an appropriate topic for your research, and assist you in locating some key reference materials. Some libraries offer more advanced tutorials or library skills workshops that help you find specific information in your area of study, teach you how to evaluate resources, help you develop critical thinking skills, and teach you more about referencing your material. Most academic libraries offer free sessions, especially near the beginning of each term. Check with the reference area to see what is available in your library.

Know how to ask for help. Walking into the reference section of a library and asking the reference staff, "How do I find stuff for a paper on obedience?" will generate a raised eyebrow for sure. Instead, determine your research topic, narrow the topic, try to develop a thesis statement, come up with some appropriate search terms, and then use an available database to search for appropriate books and articles. Now, if you are having difficulty, approach the reference staff with specific questions. For example, "I am writing a 10-page paper on obedience to authority for my social psychology class. I have located Milgram's early articles that outline an experimental approach to obedience. I would like to compare Milgram's early experimental procedures with ones that would pass an ethical review board today. Can you advise me on how I can best locate contemporary articles dealing with experimental procedures for inducing compliance in participants?"

Develop search terms and phrases. The comprehensiveness of the articles and books you find on a topic is directly impacted by the type of search terms you use. Librarians sometimes use the term "controlled vocabulary" to encompass the words, phrases, index terms, and other descriptors that help you obtain precision in database searching (Walker & Janes, 1999). Start by examining subject indexes in books on the topic of interest or consult a thesaurus to help you come up with a set of similar terms. In addition, if you already have an article on the subject, look at the key words it has listed for its journal (they can usually be found right after the abstract).

Evaluate the materials you find and be selective in your use of them. Where possible, choose scholarly articles over popular articles. Scholarly articles include terminology specific to an area of study and are generally written for use by other scholars in that area. Peer-reviewed journal articles tend to follow standardized formats such as APA style, include technical information such as statistical results based on scientific research, and cite the most relevant references. Popular articles, in contrast, are written for a general audience and tend to include opinions and commentary that may not be verifiable or supported by actual research.

Consult a guide on how to use libraries to learn more about how to retrieve important information. This chapter provides an introductory glimpse into the process for finding scholarly books and articles in the library. Lane, Chisholm, and Mateer's (2002) *Techniques for Student Research: A Comprehensive Guide to Using the Library* is a great source for developing library search skills and finding answers to questions using reference works, bibliographies, indexes, directories, dictionaries, almanacs, encyclopedias, maps, and atlases.

■ USING THE INTERNET TO LOCATE BOOKS AND ARTICLES

The World Wide Web (WWW or web) is a linkage of information on the Internet that takes a variety of digital formats, including text, pictures, audio, and video. Web browsers are computer software programs such as Internet Explorer (http://www.microsoft.com) or Netscape (http://www.netscape.com/). Browsers enable you to access and explore the World Wide Web (Ackermann & Hartman, 2000). The Web contains information on everything you can conceive of, including news, travel, entertainment, people, statistics, organizations, and all matters of interest within the social sciences. After connecting to the Internet and starting the web browser program, you can begin a search using a search tool in the form of a subject directory or a search engine.

Yahoo! is a subject directory that organizes websites within a hierarchy of categories (McGuire, Stilborne, McAdams, & Hyatt, 2000). You can conduct a search using Yahoo! (http://www.yahoo.com) by starting with the selection of a broad search category and then choosing a subcategory, followed by another subcategory of the previous one, and continue in this fashion until you find something of interest. Alternatively, you can type a key word into the search space on the Yahoo! home page to explore the entire collection of web pages that contain that key word (McGuire, Stilborne, McAdams, & Hyatt, 2000). For example, if you type in "obedience to authority," the first item that comes up in over a million matches is a book by Milgram available for purchase from Amazon.com. You can select a highly relevant match or you can select *more from this site* to find additional selections that are compatible with a relevant match.

Google is a search engine that consists of software that travels the Web and returns an index of page links ranked according to relevance (Berkman, 2000). An advanced scholar search engine is available using "Google scholar," found at http://www.scholar.google.com. Google scholar specifically helps you locate books and peer-reviewed articles along with other forms of scholarly information. With the search phrase tried on Yahoo! (i.e., "obedience to authority"), a Google scholar search produces just over 7700 matches, beginning with Milgram's (1974) book linked to a library search tool that allows you to enter your location to find the nearest library with this holding. The first 10 matches include scholarly books and citations that deal directly with obedience to authority (and in most cases, the matches bear a direct relation to Milgram's early experiments).

■ INTERNET TUTORIAL SITES

One of the best ways to become familiar with the Internet in a user-friendly fashion is to examine Internet tutorial sites. The University of California at Berkeley has a great introductory site that helps you utilize the web for scholarly resources.

Finding Information on the Internet: A Tutorial
http://www.lib.berkeley.edu/TeachingLib/Guides/Internet

You can also check out the University at Albany's Internet Tutorials site.
http://library.albany.edu/internet

■ EVALUATING INTERNET MATERIALS

The most important consideration regarding the use of the Internet for class assignments is quality. Make sure you assess the quality of the information you find on the Web. Grant MacEwan's Learning Resource Centre offers the following advice as a guide to use of Internet sources:

Is It Any Good? Evaluating Information on the World Wide Web

The World Wide Web has become a major vehicle for both finding and publishing information electronically. There is no quality control on the WWW, resulting in a wide range of quality and expertise. Because anyone can "publish" on the Web, and because so much information is available, it is necessary to develop skills to evaluate what you find. Evaluating anything you read, whether print or Web-based, involves questioning the following aspects of the source: *accuracy, authority, objectivity, currency, comprehensiveness of coverage,* and, to some extent, *ease of use.* Questionable resources are as readily available as quality information. Web pages may be sponsored by educational institutions, governments, businesses, political organizations, lobby groups, private or professional associations, and even individuals. You must ask the questions below to assess their value. The greater the number of *yes* answers to the questions listed below, the more likely the Web source is of high quality.

Authority
- Is it clear who is sponsoring the page or what institution/organization the author is affiliated with?

- Is there a link describing the purpose of the sponsoring organization?
- Is this organization recognized in the field in which you are studying?
- Is it clear who wrote the material, and what the author's qualifications are?
- Is there an address to contact for more information?
- If the material is protected by copyright, is the name of the copyright holder given?

Accuracy

- Are sources of any factual information listed in a clear and complete manner so that they can be verified if necessary?
- Is the information free of grammatical, spelling, and other typographical errors?
- If statistical data is presented in graph or chart form, is it clearly labeled and easy to read?

Objectivity

- Is the information provided as a public service?
- Is the information free of advertising, or if there is any advertising on the page, is it clearly differentiated from the informational content?
- Are the organization's biases (if any) clearly stated?

Currency

- Are there dates on the page to indicate
  - when the page was written?
  - when the page was first placed on the Web?
  - when the page was last revised?
- Are there any other indications that the material is kept current?

Coverage

- Is there an indication that the page has been completed, and is not under construction?
- If external links are provided, are they evaluated, and do they relate to the purpose of the website?
- Are any references to other sources cited correctly?
- If there is a print equivalent to the web page, is there a clear indication of whether the entire work is available on the Web or only parts of it?

Ease of Use

- Is it usually easy to connect to the site? Is it relatively free of technical difficulties?
- Is it attractive with clearly marked internal links and appropriate graphics?
- Are there tools, such as a help screen, site map, or internal search engine to guide you around the site?

Reprinted with permission from Grant MacEwan's Learning Resource Centre in Edmonton, Alberta, Canada.

CHAPTER 5

Referencing Using APA Format

If you are writing a paper or research report for a specific course, it is important that you check with your instructor in case he or she has a required citation or reference format that you need to follow. You may be docked marks if your paper does not follow the correct format. The most commonly used style in the Social Sciences is known as APA format, as outlined by the American Psychological Association (2001). These guidelines are available in an APA manual shelved in the reference section of virtually all libraries. Follow this format in the absence of any other directive. APA format is a set of rules about how to cite sources within the text of your paper or research report as well as how to list the references at the end of your work. This chapter provides an overview of APA format.

■ CITING WITHIN YOUR DOCUMENT

When you make reference to someone's work, including when you paraphrase ideas within your paper, the APA style advises you to *cite the last name of the author followed by the year of publication within parentheses.*

Akers (1991) suggests that independent indicators of self-control are necessary to clarify the relationship between low self-control and criminal and analogous outcomes.

Independent indicators of self-control are necessary to clarify the relationship between low self-control and criminal and analogous outcomes (Akers, 1991).

In the case of *multiple authors*, type "and" between their last names within the text, or use the "&" symbol if the citation is at the end of a section in your paper as follows:

> Keane, Maxim, and Teevan's (1993) earlier study on the relationship between self-control and impaired driving was the first test of the General Theory of Crime to employ a direct behavioral measure of low self-control.

> The first test of the General Theory of Crime to employ a direct behavioral measure of low self-control dealt with impaired driving (Keane, Maxim, & Teevan, 1993).

Page numbers are generally not included in your citations unless you use a direct quote.

> Gottfredson and Hirschi (1990) define crimes as acts which are "undertaken in the pursuit of self-interest" (p. 15).

> Crime consists of acts "undertaken in the pursuit of self-interest" (Gottfredson & Hirschi, 1990: 15).

If you are citing *two or more works by different authors*, list them in alphabetical order, by the first surname of the first author.

> There is debate over whether self-control is unidimensional or multidimensional (Brownfield & Sorenson, 1993; Forde & Kennedy, 1997; Grasmick, Tittle, Bursik, & Arneklev, 1993).

Personal communication. To cite information you obtained during an interview, in a lecture, over e-mail, or in any other form of personal communication, list the author's initials and last name, followed by the reference to a personal communication and the date on which the communication took place. The citation occurs within the text of your paper, but it is not included in your reference list.

> An anthropology instructor explained that "symbolic communication is not limited to humans, as evident in various alarm calls depicting different predators made by African monkeys" (L. Mutch, personal communication, March 21, 2005).

■ LIST OF REFERENCES

APA referencing requires that you create a list at the end of your documents that includes the name, title, author, date and place of publication, and publisher for each source cited in your paper. The list should be double-spaced and arranged in alphabetical order. The first line of each reference should be flush with the left margin, and any additional lines for the same entry should be indented five spaces (one-half inch). This is called a hanging indent. Most of the information you need to include in your reference list from the original source is located within the first couple of pages of the book or on the first page of the article. The exact style can differ slightly, depending on the number or form of authorship (e.g., single or multiple authors), the type of source (e.g., book, edited book, or article), and the location of the source (e.g., found on the Internet).

■ GENERAL RULES FOR LISTING AUTHORS

Single author. List the author's last name, followed by the first and middle initials of other names.

> Smith, J. A. (1984).

Two or more authors. List the authors in the order in which they appear on the title page for that work. For each person, include the author's last name, followed by the first and middle initials of other names, separated by a comma, and an "&" symbol before the last author.

> Smith, J. A., & Court, B. F. (1984).
> Smith, J. A., Court, B. F., & Baxter, J. (2004).

Multiple works by the same author. In cases where you refer to multiple works by the same author, list the references by date, beginning with the most current as follows.

> Milgram, S. (1976).
> Milgram, S. (1974).
> Milgram, S. (1969).

The reference continues to include the information needed for a book, article, or whatever the type of reference source, as indicated below.

■ TYPES OF REFERENCES

Books. Include the author(s) in the above format, the year of publication in brackets, a period, the title of the book in italics, another period, the place of publication (usually a major city) followed by a colon, and the publisher. End with a period. The book title should be italicized, and if the book has been reprinted, the title should be followed by parentheses containing the relevant edition number (e.g., 2nd) and the word "ed."

> Powell, R. A., MacDonald, S. E., & Symbaluk, D. G. (2005). *Introduction to Learning and Behavior* (2nd ed.). Pacific Grove: Wadsworth.

Edited books. List the name of the editor first, in place of an author, and follow it with "(Ed.)" Then list the year of publication in brackets, include the title of the book in italics, a period, the place of publication followed by a colon, and the publisher. End with a period.

> Linden, R. (Ed.). (2004). *Criminology in Canada* (5th ed.). Toronto: Thomson Nelson.

> Peplau, L. A., Sears, D.O., Taylor, S. F., & Freedman, J. L. (Eds.). (1988). *Readings in Social Psychology* (2nd ed.). New Jersey: Prentice Hall.

Book chapters. List the chapter author(s), the year of publication in parentheses, a period, and the chapter title followed by a period. Continue with the book editor(s)' or author(s)' names, the book's title, edition, pages that the chapter spans, place of publication and the publisher. End with a period. Note that in this case, the editor(s)' name(s) are written in full in the normal order, preceded by the word "In" and followed by "(Ed.)" or "(Eds.)" and a comma.

> Symbaluk, D. G., & Cameron, J. (2000). The warm-cold study: A classroom demonstration of impression formation. In Mark E. Ware and David E. Johnson (Eds.), *Handbook of Demonstrations and Activities in the Teaching of Psychology* (2nd ed., pp. 170–172). New Jersey: Lawrence Erlbaum Associates.

> Symbaluk, D. G. (1996). The effects of food restriction and training on male athletes. In W. David Pierce and W. Frank Epling (Eds.), *Activity Anorexia: Theory, Research and Treatment* (pp. 147–157). New Jersey: Lawrence Erlbaum.

Journal articles. For journal articles, begin with the author(s)' name(s) as discussed above, along with the date of publication in parentheses and a period. Next, include the title of the article followed by a period. List the journal title and number in italics, a comma, and the page numbers associated with the article. End with a period.

> Howell, A. J. & Symbaluk, D. G. (2001). Published student ratings: Reconciling the views of students and faculty. *Journal of Educational Psychology, 93,* 790–796.

If the journal article was reproduced online, the bibliography should indicate that it was an electronic version after the title.

> Howell, A. J. & Symbaluk, D. G. (2001). Published student ratings: Reconciling the views of students and faculty. (Electronic version). *Journal of Educational Psychology, 93,* 790–796.

If the article was obtained from an online journal, the retrieval date and URL will replace the volume and page numbers.

Howell, A. J. & Symbaluk, D. G. (2001). Published student ratings: Reconciling the views of students and faculty. *Journal of Online Psychology*. Retrieved November 18, 2005, from http://www.artsci. gmcc.ab.ca/people/symbalukd

Magazine articles. Include the author's surname and first initials, the year of publication and month in parentheses, and a period. Add the title of the magazine article followed by a period. Now put the name of the magazine in italics followed by a comma. Include the corresponding volume number in italics, the issue number in parentheses, a comma, and relevant page numbers. End with a period.

Hoffman, J. (2005, April). Male Bonding. *Today's Parent, 22* (3), 146–150.

Newspaper articles. Start with the author's last name and first initials followed by a period. Next, insert the year of publication, a comma, the month and day in parentheses, and end with a period. The title of the newspaper article comes next with a period. Now place the name of the newspaper in italics followed by a comma. Use a "p." to indicate "page," list the section with the page number, and end with a period. If the author of the article is unknown, start with the title of the newspaper article followed by the year of publication and month as indicated below.

Diotte, K. (1998, November 8). Johnny on the spot. *The Edmonton Sun,* p. SE12.

Johnny on the spot. (1998, November 8). *The Edmonton Sun,* p. SE12.

Internet sources. Include the author(s) and the year of publication in parentheses (if known), followed by a period. Next, place the title of the book, journal, or newspaper in italics (if relevant), insert the word "Retrieved" and then enter the month, day, and year that you downloaded or utilized that source along with the URL (internet address).

Lowman, J. (2004, July 20). Reconvening the federal committee on prostitution law reform. In *Canadian Medical Association Journal.* Retrieved March 21, 2005, from http://www.cmaj.ca/cgi/ content/full/171/2/147

Statistics Canada (2005). Divorces by provinces and territories. Retrieved March 25, 2005, from http://www.statcan.ca/english/Pgdb/ famil02.htm.

Reference Checklist:
- Last page of your document, numbered (in accordance with the rest of the paper)
- Centred title "References"

- Double-spaced
- Arranged alphabetically
- First line of each entry is flush with left margin; additional lines are indented five spaces (a "hanging indent")
- Leave one space after and between punctuation marks
- For each entry:
  - author(s) listed last name first, followed by initials
  - year of publication in parentheses, followed by a period
  - title of the work (capitalize each word in a book, the first word in a journal article, and each word in a journal name), followed by a period
  - volume or issue number if applicable, followed by a comma
  - page numbers if applicable
  - place of publication, followed by a colon
  - publisher, followed by a period

■ OTHER SOURCES

For more advice on how to complete your references, including how to write references based on other types of sources, such as government documents, doctoral dissertations, and works of several volumes, check out *A Guide for Writing Research Papers APA Style* online at http://webster.communet.edu/apa.

Finally, the American Psychological Association (2005) has recently published a "pocket version" of the APA Manual that includes all of the main rules needed for writing and formatting articles, along with suggestions for avoiding common grammatical errors. As a student, you might find this to be a more economical purchase than the larger manual.

■ REFERENCE MANAGEMENT SOFTWARE

There are several commercial software products available that help you collect, format, sort, search, and reuse a bibliography of sources. They include *EndNote, ProCite,* and *Reference Manager.* Most likely, you are able to purchase these programs for discounted student rates through your university bookstore or computing services. Furthermore, some universities offer their students free access to a Web-based version such as *RefWorks.*

These programs typically interact well with word processing software, such as Microsoft Word, allowing you to insert references "on the fly." Furthermore, the list of references is automatically generated for you at the end of your paper, using the style you select (such as APA). As an undergraduate student, you may find this type of software too costly or simply distracting. However, if you are writing a substantial honours project paper, or beginning a graduate-level degree, you should consider the investment.

CHAPTER 6

Writing Papers

Many courses include an essay assignment in order to facilitate critical thinking and to help students develop their writing and communication skills. Sometimes you can choose your own topic, and other times your instructor will offer you some choices or assign a specific topic. The goal of a particular essay (sometimes called a research paper) might be to summarize an area or idea, compare two or more theories, explain a view or perspective and/or evaluate the work of others. The essay writing process includes developing a thesis statement, outlining your ideas, locating supporting information, and structuring your essay in a logical fashion. This chapter describes common essay formats and provides advice on how to narrow a topic, create a thesis statement, and write a paper.

■ TYPES OF ESSAYS

An essay assignment can vary considerably in length and type depending on the course and instructor. An explanatory (also known as expository) essay usually informs the reader about a specific topic or issue. This can be accomplished with factual research, as in *Prostitution Offender Programs as an Educational Alternative to Criminal Charges*. You can also describe some phenomenon using one or more theoretical perspectives, as in *A Jungian Approach to Dream Interpretation* or *Heaven's Gate: A Multi-theory Analysis*. In either case, the reader learns about the topic in considerable detail. If you would like to learn more about Heaven's Gate and read a student's explanatory essay, please refer to Appendix 1.

An argumentative essay takes a stance on an issue and provides supportive evidence. For example, *Capital Punishment is not a Deterrent to Violent*

Chapter 6: Writing Papers

Crime or *Children with Attention Deficit Disorder Learn Best in Highly Structured Environments*. The issue can be quite controversial, as in *Affirmative Action Policies Constitute a Form of Discrimination*. Your goal in this kind of essay is to persuade your reader that your position or point of view is valid.

An essay assignment can also take the form of a book report. The objective of a book report is usually to summarize the book's main purpose and ideas. A book report may be descriptive in nature and follow the same general format as an explanatory essay. Alternatively, you may be asked to evaluate the book or write a book review. In this case, the report also renders a judgment as to whether the author(s) accomplished the intended purpose and adequately addressed the main issues. You might also be expected to indicate if you would recommend the book to others and to note its overall merit relative to other books in that area of study.

■ DEVELOPING A THESIS STATEMENT

Where do I start? Writing an essay begins with the development of a thesis statement. A thesis statement specifies the focus of your essay by summarizing or introducing your main argument, the position you are attempting to defend, or the area of interest you are going to explain. Try the following four-step process to help you develop a thesis statement.

1. Identify an area of interest.
2. Develop search terms.
3. Investigate the literature.
4. Write your thesis statement.

General	Narrow	Specific
personality	sleep problems	The influence of circadian rhythms on sleep and resulting sleep problems.
memory	repressed memories of abuse	The role of hypnosis in creating false memories of abuse.
social stratification	slavery as a form of stratification	Conditions of slavery in Canada in the 19th century.
sociological theory	symbolic interactionism	Symbolic interactionism: Comparing contributions of George Herbert Mead and Erving Goffman.

Identify the area of interest. First, identify the general area of interest for your essay (e.g., deviance, criminal justice, young offenders). Next, *narrow the topic.* If your general area was deviance, you might narrow the focus to smoking among adolescents. A specific essay topic in this area might be *Changes in Perceptions of Smoking among Adolescents.* See the table on page 44 for more examples of how to narrow a topic.

The idea here is to narrow your topic to a specific area of research. Once you accomplish this, you start to look for the most current, relevant, and supporting research on that topic.

Develop search terms by listing the key terms, phrases, and areas of research pertinent to your topic. Suppose you wanted to explore *The negative impact of circadian rhythms on sleep.* In this case, you might link *circadian rhythms* and *sleep disorders* as key terms in your search strategy. If your essay is on *The role of hypnosis in creating false memories of abuse,* key terms include *hypnosis* and *false memories of abuse.* Now it's time to see what your search terms produce.

Take a look at the literature. After developing some search terms, you need to think about gathering appropriate source materials for your paper. Rarely could you ever narrow a topic to the point where no literature exists on it. However, you may become interested in a slightly different focus when you look at the literature representing your key terms. For instance, you may want to examine *The role of hypnosis in creating false memories of abuse,* but may find a lot of articles directed at a debate about whether hypnosis *uncovers repressed memories of abuse* or *produces false memories of abuse.* Now you may wish to modify your topic to reflect the overall debate, or, at the very least, you will need to acknowledge and summarize the debate before advocating one side of it.

Write your thesis statement. A thesis is a theory, hypothesis, view, or proposal that is developed and supported in the body of the essay. A thesis statement sets the foundation for your paper and actually limits the topic in a manner that makes it manageable given the page constraints. *This paper will explore the debate on memories of abuse, and it will present facts supporting the argument that recovered memories of abuse are largely false memories produced through hypnosis and other forms of suggestion.* Once you have your thesis statement in its final form, you can continue to search the literature and gather the materials you need to begin writing your essay. The idea is to write your essay in a manner that supports your thesis. For advice on how to locate and retrieve information from the library and Internet, refer to Chapter 4.

■ ESSAY FORMAT

An essay should include the following components:

1. Title Page
2. Introduction
3. Body
4. Conclusion
5. References

■ TITLE PAGE

A title page is the first page of your essay. A title page is created on an un-numbered blank sheet of white paper that is omitted in the page count for an assignment. Try to develop a title that depicts your thesis statement, sets the tone for your essay, and gives direction to your paper. For example, you might call an essay *Limitations of Gottfredson and Hirschi's General Theory of Crime* or *Comparing Karl Marx and Max Weber's Views on the Rationalization of Society*. Centre, double-space, and bold the title, and place it just shy of half way down the page. In addition to conveying the focus of your paper, a title page identifies and classifies your work by noting the name of the author (i.e. your full name), an identification number (i.e., your student ID number), the person to whom the paper is submitted (i.e., your instructor's name), the course for which the paper is written, and the date of submission. These items are generally located near the right-hand bottom corner of the page as follows:

> Submitted by: Jeff Baxter
> Student ID: 448855991
> Submitted to: Dr. Wong
> Course: Psychology 104
> Date Submitted: April 8, 2007

■ INTRODUCTION

An introduction begins with a broad statement that identifies the area of research and gets the reader's attention. For example, *Crime affects all of us, directly or indirectly.* Sometimes, the first sentence of an introduction poses a question such as *Can all forms of crime be adequately explained by one general theory?* Another technique is to include a brief quote to set the tone and identify the issue. *Michael Gottfredson and Travis Hirschi (1990) devised a General Theory of Crime which defines crime as "acts of force or fraud undertaken in the pursuit of self-interest" (p. 15).*

Narrow the focus as you describe the theory or establish the literature around this topic. *The General Theory of Crime claims that criminal acts, for the most part, result in little or no long-term gain. Individuals commit crime because they lack self-control.* The introduction sets up the context for your essay and ends with a thesis statement that implies the structure for the remainder of the essay. *The purpose of the present study is to assess the applicability and predictive power of low self-control in Michael Gottfredson and Travis Hirschi's (1990) General Theory of Crime for explaining sexual offending.*

■ BODY

The body of your essay generally includes two or three main ideas with corresponding supporting arguments. Develop a structured outline of phrases or sentences that depict precisely what you want to convey in your essay (as demonstrated below). Begin with your thesis statement, write a main idea, and note the examples you plan to develop to support your thesis. Next, include a second main idea with supporting examples, and continue in this fashion until you have adequately summarized the literature and have enough ideas to create an essay of appropriate length (usually about three main ideas translate into a paper that is about five pages, or 1250 words). End with a reiteration of the thesis statement (worded somewhat differently) or a conclusion that answers the challenge set out in the assignment.

You may find it useful to first organize your ideas and thoughts in a point-form outline. Write your ideas on a separate piece of paper or in a new computer document. As you write, you can either refer back to the outline as a guide or simply type in sentences and paragraphs to fill out around it. Your outline might look like this:

- Introduction: Gottfredson and Hirschi's (1990) general theory of crime
- Thesis statement: Low self-control as the underlying cause of crime.
- Main point #1: Versatility in criminal and analogous outcomes
  - Supporting example: information on multiple offences
  - Supporting example: smoking, drinking, and gambling
- Main point #2: Stability over time
  - Supporting example: early criminal tendencies
  - Supporting example: recidivism following incarceration
- Main point #3: Empirical Tests of the theory
  - Supporting example: application to white collar crime
  - Supporting example: application to sexual offending
- Conclusion: Low-self control is inadequate for explaining all forms of criminal offending.

Chapter 6: Writing Papers

Use transition sentences to connect main ideas, supporting arguments, and paragraphs. A colleague of mine says she uses climbing a staircase as an analogy for explaining the need for transitions in an essay. If you are standing on the landing of a staircase (the introduction), look up the stairs to determine where you need to go to reach the top of the staircase (the conclusion). Each step on the staircase can be considered a paragraph. Step by step you work your way to the top (or end). Each paragraph (and step) is connected to the proceeding one and to the next one. Try to connect paragraphs using transitional words and phrases such as *in addition, moreover, on the other hand,* or *finally.*

■ CONCLUSION

At the end of your essay, summarize (or conclude) by noting where you started and where you ended up. In contrast to the introduction that began with a general statement that narrowed in focus, the conclusion begins with a narrow statement that is brought back into the general literature. For example, the conclusion may begin by noting whether or not a main hypothesis was supported. The final statement in your conclusion is often a reiteration of the thesis statement, a judgment, or a direction for future research.

> Although self-control was not a strong predictor of crime, it has some potential to further our understanding of sex offending. Rapists have lower self-control relative to incest offenders, who in turn have lower self-control than pedophiles. Variations in self-control within criminal groups are not accounted for in Gottfredson and Hirschi's theory, which is limited to a distinction between criminals (with low self-control) and non-criminals (with high self-control). The General Theory of Crime needs to be re-specified to account for variations in self-control within criminals (adapted from Symbaluk's 1997 unpublished dissertation).

■ REFERENCES

A list of references comprises the last page of your essay. Begin on a new page, with the centred title "References" and using the format outlined by the most recent American Psychological Association Manual (APA), list in alphabetical order all of the authors' work that is quoted, described, paraphrased or otherwise utilized in your essay. Each source includes the date of publication and the last name (in the order they appear in the original publication) of each author, separated by a comma as well as the page number in the case of an exact quote. See Chapter 5 on referencing using

APA format for step-by-step instructions on and many examples of how to set up your references.

Paper Checklist:

- Separate title page, with title and personal information
- Introduction, starting on page 1
- Body
- Conclusion
- References, starting on a new page

■ OTHER RECOMMENDED DO'S AND DON'TS

- *Consider improving your overall essay-writing skills by reviewing the basics of grammar, syntax, and language use.* Perhaps you have forgotten some of the rules of language from your early English classes? A common error in syntax I frequently encounter in student papers involves mixing up *who* and *that. Who* corresponds to people, *that* is used in reference to things. *The researcher who* utilized an experimental method versus *an experiment that* took place in 1989. I highly recommend Otte Rosenkrantz's (in press) *Right Your Wrongs* as a user-friendly reference source for avoiding common errors in syntax. It also contains all kinds of learning tips for improving your writing and provides grammar help through a review of the basics.
- *Begin the essay process early on.* Writing takes a lot of time, as does locating, reading, and interpreting the relevant literature. If you wish to earn a very high grade on the essay, plan to invest a considerable amount of time.
- *Aim for the page limit.* Assume 250 words a page, double-space, and choose a 12-point standard font such as Times New Roman. Fewer pages are clearly insufficient and more may demonstrate that you cannot write to the point.
- *Write in the proper tense.* Usually, a term paper is written in the *present tense* unless you are describing the research of others. *Bard et al. (1987) examined the rapist-child molester dichotomy in more detail with a sample of 187 'sexually dangerous' offenders from a Massachusetts Treatment Centre.*
- *Type your report and use a spell check.* Presentation counts. Even if you have great ideas, they are likely to be overlooked and under-rated if your teacher finds a number of spelling mistakes.
- *Ask a friend to read the final draft.* An independent reader is better able to tell you if your thesis statement is clear, if your main arguments appear to be well supported, and if your ideas flow in a

logical fashion. Your friend might also alert you to a few grammatical errors and spelling mistakes your computer software has overlooked.

- *Don't label your sections.* Your paper should flow naturally from the "introduction" to the "body" to the "conclusion." If your paper is quite long (over 10 pages), you can include a few subheadings to enhance the overall appearance, but be sure to also include appropriate transitional phrases.

- *Try not to over-rely on the Internet for your reference sources.* You need to make at least one trip to an actual library. Try to vary your citations using primarily books and peer-reviewed journal articles along with a few other sources (e.g., internet articles, personal communications, and government documents). Many academic institutions offer online access to numerous journals, so check out your library's website once you identify some journal articles.

- *Do not plagiarize the work of others.* If you borrow the ideas of others, you need to cite the source. There are all kinds of rules about plagiarism, and the penalty for violation can be as severe as expulsion from your program or place of study.

- *Avoid using long or multiple quotations.* The reader would rather know your thoughts than those of a cited source. Consider paraphrasing ideas you agree with or points you wish to make that originate with others.

- *Use of swear words, slang expressions, or other jargon is unacceptable.* The main character is "upset," is a better choice of wording than "pissed off." An essay is a professional document, so it is important to keep the language as such.

- *Avoid personal references.* "This paper argues" is a more appropriate way to state something than "I believe ..."

- *Steer clear of personal examples.* Refrain from telling your instructor about your past experiences and those of people you know. The support for your thesis statement should come from empirical research found in peer-reviewed journal articles and other scholarly works.

CHAPTER 7

A Research Proposal

A research proposal is generally a requirement in a third-year psychology or sociology methods course because it teaches students how to investigate an area of interest, design a research methodology, and communicate ideas through a detailed report. A research proposal is the starting point for what may later become an honours thesis or an advanced independent study project. The proposal includes a working title, an introduction, a methods section, and an anticipated outcome section. Since it includes all the main research components except for the collection, analysis, and discussion of actual data, a research proposal may be sent to a research ethics review committee for consideration before the study commences. This chapter describes the main parts of a research proposal and outlines ethical concerns involving research with human participants.

■ PARTS OF A RESEARCH PROPOSAL

Title Page. The first page of your research proposal is the title page. Centre and double-space your title, which should be about 10 to 12 words long, and locate it about halfway down the page. Come up with a tentative or working title that best describes what you anticipate finding or exploring in the proposed study. Choose a title that gives the reader a sense of the central topic and research question. For example, you might be interested in finding out if and how the portrayal of gender roles in the media has changed over the last five years. Your working title could be *Recent Changes in the Portrayal of Gender Roles on Prime Time Television.* If your subsequent

NEL   *Chapter 7: A Research Proposal*   **51** research actually indicates that little or nothing has changed in the portrayal of gender roles, you can slightly modify the title in an eventual research report to reflect your findings.

The title page is an identification source for your work, and it includes your name, your student identification number, the course name for which the proposal is assigned (the name of the instructor to whom you will give the proposal), and the due date of the proposal. These items are usually located near the bottom of the title page on the right-hand side.

Title Page Checklist:

- Separate title page, numbered
- Working title, centred, starting about half-way down the page
- Several blank lines
- Right-justified near the bottom of the page
  - Author's name
  - Student identification number
  - Course name and/or instructor's name
  - Date

Introduction. The introduction section should be double-spaced and numbered page 2, with the title "Introduction" centred at the top of the page. This section begins with a general statement that identifies your area of interest. For instance, *Canadians spend a lot of time in front of a television.* Try to narrow the focus over the next couple of sentences and end the paragraph with the key statement you wish to investigate. For example, *A recent report by John Doe indicates that an average Canadian watches three hours of television a day and, during this time, views as many as 60 advertisements. Concerns centre around what kind of messages are sent via television programming since television content is largely determined by advertising profits. Television commercials play an influential role in conveying images about males and females to the general public. This study examines the portrayal of gender roles in commercials shown during prime time television ...*

An introduction also summarizes the relevant literature. For example, *A recent meta-analysis on the content of television commercials shown in Canada indicates that men and women are portrayed very differently when they are depicted as central characters within a commercial (Doe & Smith, 2005). Women are much more likely to be shown in the traditionally feminine role of primary caregiver while men are more likely to be depicted as the main provider in an occupational role outside of the house.*

It is also important to discuss and integrate the main issues and findings within a theoretical context. For instance, *A feminist perspective emphasizes the influence of patriarchy or male dominance in advertising, pointing out that most producers are successful white males in their fifties and most central characters in television advertising are males engaged in stereotypically masculine roles (Doe, 2003).*

Lastly, this section notes the contributions of the present study to the existing literature and introduces your working hypothesis. In this case, you note how this study adds to, clarifies, or expands upon on existing research. *This study contributes to our current knowledge through an examination of gender representation by the central figure, location, setting, and product type of television advertisements.* A working hypothesis is a statement about the relationship between variables. It provides a prediction of what the current study might determine. *This study predicts that primetime television advertising continues to over-represent males (and under-represent females) in the central characters, but to a lesser extent than it did ten years ago. Finally, this study predicts that the sex of the central character is a stronger indicator of gender-stereotyped differences than location, setting, and product-type.*

Introduction Checklist:
- Starts a separate page, numbered as page 2
- "Introduction," centred
- Double-spaced
- Identifies the area of interest
- Defines the main concepts
- Summarizes key research in the area
- Establishes the theoretical context
- Introduces a working hypothesis
- Notes the contributions of the present study

Methods. The methods section details the proposed unit of analysis, setting and materials, procedures, and dependent variables for your study. Subheadings are often used to separate these sections. Begin with your proposed *unit of analysis.* If you are conducting a content analysis of gender representation in television programming as depicted above, your proposed unit of analysis might be a random sample of commercials that contain central characters who are human. If you are interested in students' views on publishing instructor evaluations, potential participants for your study can be students in an introductory psychology class at a university who will complete a questionnaire during class time.

Setting and materials. After describing your unit of analysis, list the setting and materials needed to carry out the study. For example, if students are your participants, the setting might be the regular classroom for an introductory psychology class, and the materials required might include a questionnaire, a consent form, and a pencil. Alternatively, the setting for an observational study of aggressive behaviour among children might include

Chapter 7: A Research Proposal

a daycare, a playground, or a more controlled environment (e.g., a play lab set up with particular toys and one-way glass for viewing). Materials for this kind of study may entail a more extensive list of items, such as particular kinds of toys, climbing apparatuses, mats, plastic slides, a coding instrument that lists all of the toys, and a stop watch.

Procedures. The procedures section explains in as much detail as possible the kind of information you plan to gather from your participants, the way you intend to carry out the study, and the rationale for choosing these particular methods. Suppose you are describing your intention to observe children at play in a daycare setting to determine what kind of toys are preferred as a function of age and gender. You would need to detail what constitutes an observational session, how toy preference is determined, and how and for how long each child is to be observed.

Perhaps you intend to have two observers assigned to each child in the daycare (so you can later look at inter-rate reliability). Each observer notes the age and gender of the designated child on a separate coding sheet. Gender is recorded as male or female, and age can be measured in estimated years to the closest half-year (e.g., 2.5 years, 3.5 years, 4.0 years). An observational period might be a two-hour block of unstructured play time from 8:30 to 10:30 a.m. Every toy in the daycare can be itemized on a coding sheet, and the observers are trained in advance of the observation session to ensure familiarity with all the toys. Observers use code sheets to record the onset of play with a particular toy and the amount of time spent using that toy. All sorts of coding rules are established ahead of time to determine how to deal with situational variants, such as the use of multiple toys, cases where a toy is put down and then immediately picked back up, or ways to record incidents when a child takes a toy away from another who did not wish to give it up.

Dependent variables. Dependent variables are those you plan to measure in your study. Preference for toys might be a dependent variable that is measured by *type of toy*, as checked off on a coding sheet constructed prior to the onset of the study, and *length of time spent with a toy*, as recorded to the nearest minute using a stop watch that commences when a child picks up a toy and ends when the toy is abandoned. Try to list and describe all of the main dependent variables for your study here.

Methods Checklist:
- Title "Methods," centred
- Double-spaced
- Participants and how they will be obtained
- Setting and Materials
- Procedures
- Dependent Variable(s)

ETHICAL CONSIDERATIONS

When the research you propose to undertake includes human or animal participants, is funded by an institute, or uses the assets of an institute (e.g., you plan to carry out the study at a university in one of the classrooms), it needs to be approved by a Research Ethical Review Board. Most universities and colleges that offer degree programs have an internal ethics review board consisting of faculty with research experience. This group of qualified researchers examines a written description of your anticipated research (such as the observational study outlined above) and makes recommendations based on whether ethical standards have been met involving respect for human dignity, free and informed consent, anonymity and confidentiality, and debriefing processes. Your institute can provide you with more details about how you go about the ethical review application process.

In writing your proposal, you need to discuss the relevant ethical concerns and describe how you plan to address them. Here are some questions to consider that may help you determine how ethical issues pertain to your study:

- How will you recruit your participants?
- What kind of information are you collecting?
- Will you ask participants any kind of sensitive or personal questions?
- How will responses be kept confidential?
- How will you maintain anonymity?
- Does participating in your study involve any risk of physical injury?
- Would anything in the procedures of your study have psychological implications for participants in your study?
- Is any kind of deception necessary in the procedures for carrying out your study?

The American Psychological Association is renowned for disseminating information about ethical principles for involving human participants in research (see APA, 1982). Bruce D. Sales and Susan Folkman's (2000) *Ethics in Research With Human Participants* outlines current ethical concerns and notes ways to plan a study to minimize ethical issues. Most Canadian Research Ethical Review Boards adhere to the Tri-Council Policy statement of 1998, *Ethical Conduct for Research Involving Human Participants* (Medical Research Council, Natural Sciences and Engineering Research Council, and Social Sciences and Humanities Research Council). See http://www.ncehr-cnerh.org/english/code_2 to access the Tri-Council Policy Statement. Outlined next are some of the major ethical issues incorporated in the policy.

Respect for Human Dignity. The most fundamental ethical concern assessed in research proposals is whether the procedures or any aspects of the proposed research have the potential to cause harm to the participants. Harm can be in the form of physical injury, for example, from any physical activity or exercise. It can also take more subtle forms that are difficult to define, measure, or even predict in advance of the study, as in the case of psychological stress or strain, and circumstances that have the potential to jeopardize aspects of one's cultural integrity. Every effort must be made in research to protect the best interests of participants.

An ethics committee considers, among other questions, whether the benefits of the research outweigh the potential risks of harm and whether participants knowingly and willingly agree to take on any potential risks. Finally, participants need to understand that their participation is voluntary and that they can withdraw their participation at any time without penalty. This method for minimizing potential harm is handled through consent.

Free and Informed Consent. All participants give informed consent prior to the onset of the study. Informed consent means that potential participants agree to participate in the study only after learning about all of the requirements and possible risks of participation. This information might be provided on a consent form that the participants read and then sign. For example, a consent form could note that students are being asked to participate in a 15-minute study on mood and violence on television. Requirements would include filling out a very short questionnaire assessing their current mood, watching a two-minute video clip that depicts a shoot-out from an old Western movie that is rated PG and is available in most video outlets, and then filling out another short questionnaire. Subjects are ensured that their participation is voluntary and that they are free to withdraw that participation at any time without penalty.

In the case of a proposed observational study of aggressive behaviour among children, consent would need to be obtained from the daycare director, the daycare staff, and the parents or legal guardians of the children (since the children are under the age at which they can legally consent to participating in such a study). The proposal would need to clearly outline the objectives of the study (i.e., the intended uses of the information gathered), how the participants would be observed (e.g., through the use of trained observers unobtrusively standing behind the one-way glass viewing area, oblivious to the children), and so on. Videotaping of children is fairly controversial as a research method since the researcher would possess copies of the children on tape. Although it increases reliability in the coding of aggressive behaviour, this would not be a recommended method of observation since it poses additional risks (e.g., privacy) to an already vulnerable participant group.

Respect for Privacy and Confidentiality. Another ethical concern related to participating in research centres on privacy, anonymity and confidentiality. Privacy is now becoming increasingly protected by legislation (e.g., Freedom of Information and Protection of Privacy Act in Alberta) and must be handled in accord with existing guidelines. Guaranteeing anonymity to your participants means that names and other identifying forms of information are not collected by the researchers (e.g., questionnaires are numbered, but participants refrain from putting their names on them). That way, a researcher cannot link the individual participant to his or her responses. Confidentiality means that although a researcher may be able to identify a given respondent (e.g., perhaps the participant in a case study was a former prostitute who is known to the researcher who interviewed her), the researcher promises not to do so publicly. In this case, the researcher may report some of the conversation shared in the interview but promises not to release any details that might identify the individual interviewee.

Use of Deception and the Need for Debriefing. In rare cases, some details of a study are initially withheld from the participants. In an experiment, for example, it may be essential to provide incomplete disclosure about the procedures in order to manipulate an independent variable that tests a causal relationship that cannot be explored with full disclosure. For instance, a researcher may wish to study students' reactions to a request for course assistance by another student who is a stranger (e.g., a fictitious student). If the researcher informed participants in advance that someone unknown to them was about to ask for their course notes, the psychological process would be compromised. In this case, the researcher explains the need for deception in the research proposal and outlines that participants would be notified of the deception and its purpose at the end of the study (e.g., immediately after the decision to lend or not lend the notes becomes apparent).

All of these sorts of considerations are weighed out in advance by an ethics review board that ultimately determines whether the study can proceed as described (i.e., it receives ethical approval), whether the procedures or other elements of the proposed study need to be modified before proceeding (i.e., conditional approval), or whether potential harm exceeds the overall benefits of the research (i.e., the study is not granted ethical approval and cannot be carried out as described). Consult with your instructor about ethics and the possible need to undergo a review process if you think your proposed research might entail important ethical considerations.

■ ANTICIPATED RESULTS

Finally, your proposal can include a short section that details what you expect to find in this study. You might suggest that based on the earlier review of the literature, you anticipate that boys are more likely to choose toys that are loud, fast, and action-based (e.g., cars, trains), while girls may prefer toys related to social roles (e.g., doctor's kit, doll and carriage). You can also try to explain the implications of your research, should the findings turn out the way you anticipate. *Despite claims that we are becoming less gender-focused in how we rear our children, males and females are still behaving in gender-stereotyped ways as evident in their choice of toys.*

■ PROSPECTIVE REFERENCES

End your proposal with a list of any materials (e.g., scholarly articles and books) you cited in the introduction. Since the study is still in a preliminary stage, students sometimes refer to this as a list of "Working References" or "Prospective References." In this case, the list can also include references you plan to include (perhaps in the discussion) or that you currently think are relevant. This helps your instructor determine your familiarity with the key theorists, research, and methodologies relevant to the proposed area of study. Please refer to Chapter 5 for a detailed description of how to properly reference the work of others.

Research Proposal Checklist:
- Title Page
- Introduction
- Methods
    - Unit of analysis
    - Setting and materials
    - Procedures
    - Dependent variables
- Ethical concerns
- Anticipated results
- Prospective References

CHAPTER 8

Writing a Research Report

The requirement to produce a research report generally has some underlying objectives. You are expected to select, examine, and evaluate an area of social research. Research involves not only conducting a study but understanding the area of interest and becoming familiar with the work done by others. Importantly, the assignment teaches you about research design by having you develop and carry out original research. Finally, the task culminates in writing a report that demonstrates what you have learned through the collection, analysis, and interpretation of actual data. This is an opportunity to try to integrate your research with the existing literature.

By following the suggestions provided below, you can write a research report that impresses your instructor, provides the structure for a possible honours thesis, helps prepare you to design more advanced research (e.g., a master's thesis), and gets you into the practice of using a format required by most scholarly journals (where you might even publish your study).

■ FORMAT

A research report is structured as follows:

1. Title Page
2. Abstract
3. Introduction
4. Methods Section
   a. Participants
   b. Procedures
   c. Design
5. Results
6. Discussion

7. References
8. Tables
9. Figures
10. Appendix (optional)

■ TITLE PAGE

Let's begin with the first page of your research report—the title page. The most obvious function of this page is to label your work. Your title page has many components including an identifier title in the top right-hand corner. This is a mini-title that you can insert as a "header" that will appear on every page of your research report. The mini-title helps to quickly identify your work and piece it back together in case the pages get dropped, or another assignment gets mixed in with yours. A page number should appear after your mini-title, usually as part of the same header. I chose the header "pain mediators" for a study I conducted that had the full title *Social Modeling, Monetary Incentives and Pain Endurance: The Role of Self-Efficacy and Pain Perception.* (See Appendix 2 for an actual research report on impression formation written by one of my students in a third-year methodology course.)

A running head is a slightly longer version of your identifier title. The running head should be between four and six words long and is presented in capital letters flush with the left margin about an inch down the page. You can also include a couple of key words below this that describe the literature or identify the main area of research (e.g., social modeling and pain perception). Key words help track your published study in electronic media and via the various library systems. Thus, anyone interested in self-efficacy, pain perception, and social modeling will be able to find the article.

A title is used to convey the main focus or idea underlying your study. Try to create a title that gives the reader a sense of what the study is about (e.g., *Affirmative Action* is too vague, while *Affirmative Action Hiring Discriminates Against Qualified White Male Police Candidates* tells us what to anticipate in the actual study). The full title is usually 10 to 12 words long and appears centred on the title page, about four double-spaced lines down.

The author line is directly underneath your title, usually one or two double spaces below, and it is sometimes prefaced with "by." The author line includes your name and the names of anyone else who was directly involved with your research (e.g., if it was part of a group project with a group mark, you would include everyone's names). Below this, note the institute with which your research is affiliated (i.e., your school's name).

Next, be sure to include the date of submission. This is usually located in the lower left hand side of the page, flush with the left margin.

If a research grant was used in any way to support your study (e.g., you paid your participants with money from your instructor's central research fund, you bought equipment to tape-record a focus group session with money from your program of study, or you applied for and received a travel grant to support a poster presentation of your findings at a conference), the source of the funding should be cited on the title page. Use a separate line, flush with the left margin, just below the date of submission.

Finally, you need to include the name of a main contact person with a mailing address so that interested parties can correspond with that person for more information about the study. (Unless someone else headed up a group project in which you participated, you would put yourself down as the main contact person.)

Title Page Checklist:
- On a separate page
- Identifier title in the top-right hand corner
- Page number in the top-right hand corner
- Running head, left margin below author line
- List of key words, left margin below running head
- Full title, centred, starting about 4 inches down the page
- Author line a few down from the full title
- Institution's affiliation below the author line
- Date submitted
- Funding source if applicable
- Corresponding author and address

■ ABSTRACT

An abstract is often the most crucial page of a research report because a reader may skim over it first to determine if an article is relevant or merits further reading. An abstract is a very brief summary of the research. It notes the main question or central hypothesis under investigation, describes the participants, outlines the design, highlights the main findings, and notes implications for future research. A research abstract should answer the following questions:
- What is the main research question?
- Who are the participants?
- What is the research design?
- What are the main findings?
- What conclusions are reached?

Although the abstract is located at the beginning of your report (after the title page, on a separate page), it is usually the last thing you actually write. When you are ready to compose the abstract, it is a good practice to list everything you want to include in it and then rework it until it is very precise and concise (i.e., less than 150 words).

Abstract Checklist:

- On a separate page
- Begins with the heading "Abstract," centred at the top of the page
- Single-paragraph (not indented)
- Double-spaced
- Short (e.g., less than 150 words)
- Numbered as page 2

■ INTRODUCTION

The heading "Introduction" is generally centred, at the top of page 3 (although in some cases, the actual title of the report may be listed here). An introduction states the purpose of your research and spurs the reader's interest in your study. This section outlines the literature and includes a statement of the central research question along with the hypotheses being tested.

Questions that will help you develop your introduction:

- What is the central research question or problem under investigation?
- What does the literature say about this area of research?
- How does the present study contribute to the literature in this area?
- What are the main hypotheses?

Start your introduction with a very general statement that establishes the area that your study relates to. For example: *Pain is a fundamental fact of life.* Continue narrowing the focus until you have described the problem under investigation and located it in the relevant literature. Also, make sure you write in the present tense throughout the introduction unless you are referring to what other researchers have found or done in their studies. For example,

> People are constantly faced with minor aches and pains due to overexertion, headaches, dental problems, and other conditions … An important question concerns the conditions that regulate endurance of pain … Research in the social psychology of pain has demonstrated that social modeling can be used to increase (or decrease) pain tolerance (e.g., Craig, 1986). Extrinsic reinforcers (e.g., monetary incentives) have also been shown to increase people's endurance of a painful event (Cabanac, 1986) … (Symbaluk, Heth, Cameron, & Pierce, 1997: 258).

Finally, describe how your study makes a contribution to the literature and note the main objective of your research. For example,

> Research on pain behaviour is well established, but few studies have focused on pain mediators. The primary goal of the present study is to

assess whether the effects of social modeling on pain endurance are mediated by self-efficacy expectancies, by pain perception, or by both of these processes (Symbaluk, Heth, Cameron, & Pierce, 1997: 258).

Be sure to state your hypotheses as precisely as you can and relate them back to the earlier literature. For example, *Consistent with Cabanac's (1986) early research, the present study predicts that people who are paid $2.00 per 20 seconds, will endure isometric sitting longer than participants who receive $1.00 or no payment in the control condition.*

Introduction Checklist:
- Follows the abstract
- Begins with the heading "Introduction," centred at the top of page 3
- Double-spaced
- Indented new paragraphs
- Format conforms to an hourglass shape

■ METHODS SECTION

This section follows the introduction (without a page break) and begins with the centred title "Methods." The methods section essentially tells the reader in considerable detail exactly what materials were needed to set up the study and how it was carried out.

Questions to ask yourself as you create your methods section:
- Who were the participants?
- What supplies were needed to carry out this study?
- Were the procedures described in a manner that is so clear and straightforward that anyone reading it would be able to replicate the study?
- Was there a precise explanation for how the variables were measured?

It is useful to use a number of subheadings in this section to identify the components you need to explain (see the checklist below). The methods section usually begins with a short paragraph that indicates how many people participated in the study, describes the participants in the research, notes how they were selected, and describes how they were assigned to conditions. For example,

One hundred and twenty-three students from an introductory sociology course at the University of Alberta volunteered to participate in a study on endurance and motivation during regular class time. The class was randomly selected from a computer-generated list of all the courses offered during that fall session. All participants completed a questionnaire designed to examine views on smoking.

The methods section also includes a paragraph that lists the setting and necessary materials used to conduct the study (e.g., the location, any printed materials, types of supplies that had to be purchased in advance, and/or any kind of equipment that was needed to record information). This information helps a reader understand exactly what would be needed to replicate a study of this nature. For example, *The experiment was conducted in a regular class room at the University of Alberta. Necessary materials included a tape player, a prerecorded cassette of two confederate voices, background information sheets, paper, and pencils.*

Next, you want to include a very detailed description of all of the procedures for carrying out the study. This section literally walks the reader through the study as the participants would have experienced it. (Again, you might consider using subheadings to ensure that you document every step undertaken in this research project.) For example, if the experimenter needed to prepare some materials in advance of the study, such as a script to ensure that all participants were given identical instructions before commencing the task, the first heading might be "Background preparation." The next heading could be called "Starting the session," and here you can summarize what the participants were told about the study and what their role would entail. For example,

> Upon arrival for the experimental session, each participant was greeted by a female researcher and taken to a small laboratory. Participants were seated at a table in front of a television monitor and asked to fill out a pre-study questionnaire and to watch one of six standardized presentations on videotape (Symbaluk, Heth, Cameron, & Pierce, 1997: 260).

The procedures section essentially details the independent variable(s) in an experiment (since this is what the research manipulates). For example,

> The videos allowed for the manipulation of social modeling. Participants were randomly assigned to one of three modeling conditions (intolerant, tolerant, and no-model control condition). Individuals in the intolerant condition saw the male confederate in the isometric position wearing a heart-rate monitor. The confederate ... acted according to a script that called for him to collapse after 60 seconds ... Cue cards were used to tell the confederate what pain estimates to call out and what signs of pain to display ... Pain was displayed by facial expressions, moans, clenched teeth and hands, and shaking legs ... A similar procedure was used to create the tolerant model condition ... but the final pain estimate was not reported until he endured 240 seconds of isometric sitting (Symbaluk, Heth, Cameron, & Pierce, 1997: 260–261).

In a report that is based on the examination of existing records (i.e., secondary data analysis), the methods section might be considerably shorter, as in this article on the effects of temperature on temper among baseball pitchers. Here, the entire methods section is described in one concise paragraph as follows:

> Microfilm issues of major daily newspapers were consulted to obtain data on weather and major league baseball games. Random samples of games were taken from three major league baseball seasons: 1986, 1987, and 1988. The 1986 sample included every 10th game played during the season (n = 215 games). Every 7th game during the season was included for the 1987 (n = 304) and 1988 (n = 307) samples. For each game sampled, the number of players hit by a pitch (HBP) was recorded. Within the same newspaper issue, the high temperature (F°) in the home city of the day of the game was also recorded. The numbers of walks, wild pitches, passed balls, errors, home runs, and fans in attendance in each game were recorded as control variables. (Reifman, Larrick, & Fein, 1999: 308–309)

The research design usually has it own main heading within the methods section. This section very succinctly describes the overall form of the study (although a reader should also be able to infer this from the previous description provided in the procedures section). For example, *The pain study was a 3 × 3 factorial experiment that crossed three levels of social modeling (no model, intolerant model, tolerant model) with three levels of payment (zero, $1.00 per 20 seconds, and $2.00 per 20 seconds of isometric sitting).*

The last subsection lists the dependent variable(s) and describes how they were measured. For example, *Dependent measures included pain threshold and pain endurance. Pain threshold was recorded as the number of seconds to the first report of pain. Pain endurance was the total number of seconds of isometric sitting.*

Methods Checklist:

- Follows the introduction
- Begins with the heading "Methods," centred
- Page numbering continues from the previous section
- Double-spaced
- Indented new paragraphs
- Includes subheadings such as
  - Participants
  - Setting and Materials
  - Procedure
  - Design
  - Dependent Variables

■ RESULTS SECTION

The results section directly follows the methods section with a centred title: "Results." Here, the type of data analysis is mentioned along with any rationale needed to justify a given statistical procedure. For example, *Analysis of variance (ANOVA) indicated a main effect of social modeling on pain endurance.* This section highlights the main findings of the study in a technical manner without elaboration (i.e., do not discuss whether hypotheses were or were not supported—that is for the discussion). For example, *Participants in the tolerant condition lasted significantly longer than individuals assigned to control and intolerant conditions ... There was no main effect of monetary payment on endurance.*

The findings are generally noted as they pertain to the dependent measure(s) and are presented in the order in which the variables were introduced in preceding sections. The American Psychological Association (APA) (2001) manual is very clear about how to present the information from specific statistical analyses (e.g., correlations, t-tests, ANOVA), so make sure you refer to the guide when you are writing this section of your report. A few examples are given below.

- **Descriptive Statistics.** *A comparison of the means showed that participants who recited the information aloud with a friend performed better on the exam ($\underline{M}$ = 83.5%, $\underline{SD}$ = 12.3.) than those who did not verbalize their notes ($\underline{M}$ = 78.8%, $\underline{SD}$ = 15.0).*
- **Correlation.** *Early trouble in school was moderately associated with job instability, $\underline{r}$ (236) = .56, $\underline{p}$ < .001, and alcohol abuse, $\underline{r}$ (236) = .60, $\underline{p}$ < .001.*
- **Analysis of Variance.** *Findings showed that athletes in the low-weight class had lower body masses ($\underline{M}$ = 20.4, $\underline{SD}$ = 1.7) than those in the high-weight class ($\underline{M}$ = 25.3, $\underline{SD}$ = 0.8), $\underline{F}$ (1, 12) = 13.31, $\underline{p}$ < .01.*

You may want to include a table that summarizes the findings by listing the means (and standard deviations) for key variables. Make sure you use a transitional sentence that refers to your table and works it into the report. For example, *As shown in Table 1, participants in the tolerant condition endured pain longer than those in control or intolerant conditions* or *Participants in the tolerant condition endured pain longer than those in control or intolerant conditions (see Table 1 below).*

Alternatively, you may wish to include a figure that displays the main results pictorially as a graph or chart. You can construct a figure using various computer programs (e.g., Microsoft Excel can produce a chart that is a pie, column or line graph). In this case, the independent (or predictor) variable is plotted on the horizontal (x) axis, while the dependent (or outcome) variable is plotted on the vertical (y) axis. Make sure you include labels so that it is clear what information is being conveyed. Again, you want to use a transition sentence that refers to the figure in the results section. For example, *Participants in the tolerant condition endured pain longer than those in the intolerant or control conditions, as shown in Figure 1 below.*

Results Checklist:
- Follows the methods section
- Begins with the heading "Results," centred
- Page numbering continues from the previous section
- Double-spaced
- Indented new paragraphs
- Main findings are reported
- Optional Table lists data in columns and rows
- Optional Figure shows the main results

■ DISCUSSION SECTION

This section follows the results with the centred title "Discussion." A discussion elaborates on the main findings and notes the implications for future research.

Discussion questions to consider:
- Did this section provide an overall summary of the study?
- Was the study situated within the relevant literature?
- Were all the key findings described?
- Are the main findings related back to the literature?
- Are there any alternative explanations for the findings?
- Is the importance of the study evident?
- Are limitations of this study introduced?
- Were suggestions provided for ways to improve this research?

Results are generally discussed as they pertain to the research questions or hypotheses posed in the introduction. In this case, each finding is discussed in some detail as you note whether the result is in accord with the original hypothesis (and thus lends further support to the area of research on which you are building), or you might speculate why you did not find what you expected. For example,

Results indicated that monetary reward had no significant effect on pain endurance ... Money may have become less salient in the context of the social modeling manipulation. On the post-experimental questionnaire, most individuals reported self-esteem or competitiveness as their motivation for participating, rather than the money (Symbaluk, Heth, Cameron, & Pierce, 1997: 266).

Finally, note any limitations of the present study and how you might improve upon this research. Perhaps you used a fairly small sample, and now you'd like to replicate the study with a larger group or a different group to see if the findings hold true. The discussion ends with a conclusion to your report that briefly sums up the study and its importance. *(The present study makes it clear that ... This study has shown that ... Future research is needed to ...)*

Discussion Checklist:
- Follows the results section
- Begins with the heading "Discussion," centred
- Page numbering continues from the previous section
- Double-spaced
- Indented new paragraphs
- Main findings are described
- Findings are situated in the relevant literature
- Implications for future research are noted

■ REFERENCES

The list of references has a centred title "References" and begins on a new page. This section is double-spaced and includes a reference to anyone else's work that is described, paraphrased, or otherwise utilized in your report. Please see Chapter 5 on how to correctly cite and list references.

References Checklist:
- Separate page
- Double-space
- APA format (unless otherwise noted by your instructor or the editor of a journal)

■ TABLES

Although you most often use a table to display your main findings in the results section, the actual table is located after the reference section of your report. Tables are placed in their own section because they sometimes detract from the text of your report. In addition, when you send an article to a publisher, the editor reserves the right to fit your table into the text of the document where it is most appropriate, given the space limitations and page setup of the actual journal. You can refer to the table in your results section by stating, *as shown in Table 1* (and then indicate that *Table 1 is about here* in the text). That gives the reader an idea of where you would want the actual table to appear if the manuscript were published as an article or book chapter. If you include two or more tables, you can include a List of Tables. A List of Tables is similar to a Table of Contents and lists a brief description of each table along with the page number for that table. The List of Tables is located on the page preceding Table 1.

Tables Checklist:
- Optional list of tables
- Each table on a separate page thereafter
- Double-spaced
- APA format (unless otherwise noted by your instructor or the editor of a journal)
- Limited amount of information
- End of your report, if there are no figures or appendices

■ FIGURES

In some cases you might find it useful to include a figure that shows the main findings in a graph, chart, or other diagram. Again, you can refer to the figure in your results section by noting, *as shown in Figure 1* (with the reference that *Figure 1 is about here*). If you incorporate two or more figures, you might include a List of Figures that introduces the actual Figures found on subsequent pages.

Figures Checklist:
- Optional list of figures
- Each figure on a separate page thereafter
- Double-spaced
- APA format (unless otherwise noted by your instructor or the editor of a journal)
- Limited amount of information
- End of your report, if there are no appendices

■ APPENDIX

The appendix is an optional section for most research reports and would not normally be counted in the total number of pages if there is a page limit on the assignment. The appendix is the place to include extra information that would be useful to the reader (or someone who might want to know more details about your study to replicate it). However, it is not essential to the actual report. For example, if you used a questionnaire to find out people's views on capital punishment, you could include the actual survey in this section. Other documents that would be appropriate for inclusion in an appendix might be a coding instrument or a set of coding rules, a script used by the researcher in an experiment, a consent form, an observation checklist, or some background information distributed to participants.

Appendix Checklist:
- Includes extra documents that might be of interest to a reader
- Begins with the heading "Appendix," centred at the top of the page
- Double-spaced

■ ADDITIONAL RESOURCES

Northey, M., & Timney, B. (2002). *Making Sense: A Student's Guide to Research and Writing*. (3rd ed.). Toronto: Oxford University Press.

Pyrczak, F., & Bruce, R. R. (1998). *Writing Empirical Research Reports*. (2nd ed.). Los Angeles: Pyrczak Publishing.

Rosnow, R. L., & Rosnow, M. (2003). *Writing Papers in Psychology: A Student Guide to Research Reports, Essays, Proposals, Posters, and Brief Reports*. (6th ed.). Belmont: Wadsworth/Thomson Learning.

CHAPTER 9

Preparing for Life after the Degree: A Curriculum Vitae

Many of the career paths available to a graduate with a Bachelor of Arts or a Bachelor of Education begin with the submission of a curriculum vitae (usually to a human resource department in response to an advertisement for a job). A curriculum vitae (CV) documents your personal contact information, educational background, work experience, and academic achievements (e.g., awards, publications). *A CV is different from a resume.* If you have ever applied for a part-time job or a summer position, you probably have a resume. A resume means "abbreviated" or "summarized" in French, and it provides employers with a short version of your education and work experience. It is typically no more than one page in length. A CV, on the other hand, means "course of life" in Latin, and it is a more comprehensive list of your accomplishments.

A CV is also a work in progress. Even when you secure a permanent position with some organization, your CV should be updated on yearly basis to reflect current achievements and ongoing development. You can expect to be asked to produce a CV on occasion throughout your career, sometimes with very little notice. For example, you may be asked to provide a supervisor or department chair with a copy for peer review or evaluation purposes. A CV is generally included in an application for promotion to an administrative position or a tenured position within your department. Many research funding agencies ask for a CV along with a grant proposal. Importantly, keeping your curriculum vitae updated helps you keep track of your teaching and research experiences, scholarly awards and accomplishments, and changes to important contact information sources, such as personal references.

You will be amazed at how and how much your academic profile builds over time. A talk you gave in an introductory psychology class on *The Subsystems of Memory* in the fall of 2006 might start a list of professional presentations. The next year you might add a Poster Session you gave on *Declarative Memory* at an annual conference of the American Psychology Association. In 2008, you might present your own research on *Semantic Memory* at your university's annual research day, and you will revise your CV to include this as your third professional presentation entry.

You never know when your dream job will appear in an advertisement or some great opportunity will arise on short notice, so be ready. *Start collecting relevant information, dates, and achievements for your CV right now!* Open a file on your computer and call it *My Curriculum Vitae.* The rest of this chapter provides you with step-by-step advice for ways to translate what you have done into a CV that demonstrates your knowledge, skills, abilities, and accomplishments.

■ THINGS TO INCLUDE IN A CV

Note that there are no hard rules about what to include in a CV or how to order these items, so feel free to modify these suggestions to suit your own style.

Contact Information and Education. The first page of your CV generally includes personal contact information and describes your educational background. Begin with the centred title "Curriculum Vitae." Beneath this, insert your full name. On the next line, place your mailing address. Underneath, include your work or home phone number, fax number if applicable, and your e-mail address. The personal contact information should stand out on your CV, so it is okay to type this in bold, italics, or some kind of special but easy-to-read style.

Curriculum Vitae

Diane G. Symbaluk
Psychology and Sociology Department
Grant MacEwan College
6-329, 10700-104 Avenue NW
Edmonton, Alberta, Canada
T5J 4S2

The next section of your CV is your educational background. List your academic accomplishments, beginning with your highest degree or most recent one. Put the year you earned the degree, flush with the left margin, followed by a space and then the title of the degree, along with the area of specialization, the educational institute that issued the degree, and a title or some other kind of information that indicates the topic of research.

Education

1997	**Doctor of Philosophy, Sociology** (Criminology)(Social Psychology) University of Alberta, Edmonton, Alberta. *An application of the General Theory of Crime to Sex Offenders*
1993	**Master of Arts, Sociology** (Experimental Social Psychology) University of Alberta, Edmonton, Alberta. *Money, Modeling and Pain: The Role of Self-Efficacy and Pain Perception.*
1991	**Bachelor of Arts with Honours (Sociology)** University of Alberta, Edmonton, Alberta. *Activity Anorexia and its Implications for Amateur Wrestlers.*

Perhaps at this point, you have a high school diploma and are working towards your Bachelor of Education degree. List the degree you are working on with an expected completion date, followed by your high school diploma.

Education

2007*	**Bachelor of Education** Elementary Specialization: K to Grade 3 Queens University, Kingston, Ontario *expected completion date
2002	**High School Diploma** Bev Facey High School, Sherwood Park, Alberta

If you have any additional certificates, awards, or acknowledgments that show you have other academic or scholarly accomplishments or skills, include them next so that they stand out to the reader. These could include technical certificates, diplomas, professional development or program completion certificates, as well as titles you have earned (e.g., Microsoft Certified Systems Engineer, Victim Services Training, Rehabilitation Practitioner).

- Basic First Aid with CPR, St. John's Ambulance, March, 2005
- Microsoft Certified Professional, NT4, January, 2005
- Academy of Leadership and Training, April, 2003
- Toastmasters International, July, 1996

Honours and Awards. If you have ever received money in the form of a grant, scholarship, award, travel subsidy, or sponsorship, include the name and a brief description of the award in a new section. Some people indicate the actual monetary value of the award and/or the applicable dates when the funding occurred. You may also include subsidized trips to another city to read an essay in a contest or to play a sport in a championship game, or an award from a church or community group. If you received the same award more than once, you can indicate the multiple years in which you received it. If you have a number of awards, consider dividing them into a few sections as shown below.

Honours and Awards

Social Sciences & Humanities Research Council Doctoral Fellowship, 1995–1997
Province of Alberta Graduate Scholarship, 1994–1995; 1992–1993
Department of Sociology Research Assistantship, 1990–1991

Research and Travel Grants

Arts and Science Divisional Research Grant, 2000
Clifford H. Skitch Travel Award, 1994; 1993
Sociology Graduate Travel Award, 1994; 1993

Work Experience. The next section of your CV details your work experience. Indicate your recent employment history by naming the places of employment, your job titles, years of employment, and the main duties or applicable skills required for the positions. If possible, list some of your individual accomplishments. All jobs have accompanying duties—what special skills did you learn that might interest a potential employer? For this section, you might want to use a format similar to the one you used for your education.

2003–present   Hostess
Earl's Restaurant, Edmonton, Alberta

- Greeting and seating customers, handling gift certificates, helping to clear tables
- Special skills: supervised staff members, designed promotional events

2001–2003   Cashier
London Drugs, Sherwood Park, Alberta

- Processing customer transactions, shelving stock, dealing with customer inquiries, electronic payment processing
- Special skills: service recovery, retail communications

Since this is an academic CV, the focus is on teaching and research. If you are an undergraduate student, you likely have no direct teaching experience and limited research experience. In this case, incorporate some of your accomplishments from classes that you have taken. For example, create a heading "Professional Presentations" and itemize class presentations you have given, along with any other talks you have given as part of your work, committee, or extra-curricular activities. List your name as the author (and include the names of others if this was part of a group presentation), the title of the talk, its purpose, and the date it occurred.

Professional Presentations

Michaels, C. "Why we need to use animals in research." *Presentation given in Experimental Psychology,* Fall, 2005.

Michaels, C., Adams, L. S. & Tate, P. "Stress: What is it and how can we manage it?" *Group Presentation for Introduction to Psychology,* Winter, 2005.

Michaels, C. "The Ice Breaker." *Short Speech at Bowmen Toastmasters Club,* Winter, 2004.

How can you get some teaching experience? As an undergraduate, consider asking a former instructor of yours if you can help proctor a final exam or provide any other course support to gain some experience. You can even note that you are trying to build your CV and are looking for activities that relate to teaching. Many university departments offer teaching assistantships to graduate students who give guest lectures, help proctor and mark exams, and tutor undergraduate students for a designated course and professor. These experiences can all be included in your CV as shown below.

Teaching

Exam proctoring for Dr. Huntz. Social Psychology (Sociology 241) Concordia College, Fall, 2007.

Teaching Assistant to Dr. Wiess-Brooks. Introductory Sociology (Sociology 100)
Department of Sociology, University of Calgary, Fall, 2007.

Guest Lecturer. Profiling Prostitution Offenders. Criminology (Sociology 225)
University of Regina, Winter, 2006.

Eventually, you can replace the guest lecture and assistantship experiences with actual teaching appointments.

Instructor: Introductory Sociology (Sociology 100). Arts and Science Division, Grant MacEwan College, Winter, 2008.

Instructor: Social Psychology (Sociology 241). Department of Sociology, University of Lethbridge, Fall, 2008.

Laboratory Instructor: Introduction to Social Statistics (Sociology 210). University of Winnipeg, Fall, 2008.

Include a separate section for your research experience. Ideally, you want to include publications that demonstrate your scholarly work. If you do not have a publication at this point in your life, you can create a list of essays and research reports. For each paper, include the title to indicate the area of research and the course name to reveal your program of study. To demonstrate your abilities, emphasize when you have conducted data analyses, carried out research, and employed different methods, as shown in the examples below.

Research

Scholarly Papers

Cheng, T. *A Jungian Approach to Dream Interpretation.* A 10-page expository essay turned in for 30% of the grade in Psychology of Consciousness, Winter, 2004.

Cheng, T. *Capital Punishment is not a Deterrent to Violent Crime.* A 15-page argumentative essay submitted in a Criminology course, Fall, 2003.

Data Analyses

- conducted data analyses using a variety of statistical packages including SPSS, Excel, and SuperAnova.

Research Methods Course (Sociology 315)

- *Survey Research:* created a questionnaire to assess people's views of street prostitution. The questionnaire included open-ended items, 5-point Likert scales, and forced choice responses.
- *Experimental Methods:* designed an experiment for testing taste preferences among cola drinkers.
- *Research Report:* using an observational method, wrote a 20-page research report on gender representation in advertising.

Publications. If you have scholarly publications, list the most recent publication first. When you become fairly well established and you have multiple publications, you can start to subdivide this section into books or manuals, book chapters, scholarly publications, and research abstracts. A sample of some of my publications is given below.

Books/Manuals

Powell, R. A., MacDonald, S. E. & Symbaluk, D. G. (2005). *Introduction to Learning and Behavior: Second Edition.* Pacific Grove: Wadsworth. (First Edition: 2002).

Symbaluk, Diane G. (2004). *Study Guide to Accompany Linden's Criminology: A Canadian Perspective.* Toronto: Harcourt Brace. (First Edition: 2000).

Symbaluk, Diane G. (2004). *Test Bank* to Accompany Kendall, Lothian-Murray, and Linden's *Sociology in Our Times: The Essentials: Second Edition.* Toronto: ITP Nelson. (First Edition: 2001).

Scholarly Publications

Howell, A. J. & Symbaluk, D. G. (2001). Published student ratings: Reconciling the views of students and faculty. *Journal of Educational Psychology*, 93, 790–796.

Symbaluk, D. G., and Cameron, J. (1998). The warm-cold study: A classroom demonstration of impression formation. *Teaching of Psychology*, 25, 287–289.

Symbaluk, D. G., Heth, C. D., Cameron, J. & Pierce, W. D. (1997). Social modeling, monetary incentives and pain endurance: The role of self-efficacy and pain perception. *Personality and Social Psychology Bulletin*, 23, 258–269.

Web-Course Tools and Web Resources

Symbaluk, Diane (2005). Chapter Resources for the companion website that accompanies Del Balso and Lewis' *First Steps: A Guide to Social Research*. (2nd Canadian Edition). Toronto: Thomson Nelson.

Symbaluk, Diane (2003). Student Resources for the companion website that accompanies Aronson, Wilson, Akert, and Fehr's *Social Psychology*. (2nd Canadian Edition). Toronto: Prentice Hall.

Committee work is a form of professional service that is expected in many areas of employment. Committees can take a multitude of forms, including voluntary organizations, planning groups, church groups, and associations. Everyone takes a turn to help their department develop new policies, hire new faculty, and operate within the larger institute. Committee membership has all sorts of benefits, including exposure to new ideas, learning about politics, and working with different people. An interviewer may take particular notice of this section of your CV because committee membership conveys the more subtle message that you are people-oriented, that you can work in a group, and/or that you get along with others.

Committee work often involves administrative tasks and functions that can teach you important skills needed for eventual promotions (e.g., from a teacher to a vice-principal, from a faculty member to a chair). Associations and other kinds of memberships also demonstrate that you have other academic interests and affiliations. Networking is often achieved through memberships in associations and these connections can also help to further your career aspirations. Committee and association members also constitute a great source for obtaining references, so consider joining one soon.

Administrative/Committee Experience

Member, Council on Student Life,
Concordia College, 2005–2006

Elected Vice-President, Sociology Graduate Students' Association,
Department of Sociology, University of Lethbridge, 2003–2004

Member, Corporate Challenge Planning Committee,
Grant MacEwan College, Winter, 2000

Secretary, Bowman Toastmasters Club,
Winnipeg, Manitoba, 1999–2000

References. The last section of your CV usually includes a current list of personal references. When building a list of people willing to provide you with a positive reference, try to cover a range of academic, work, and personal affiliations. At least two references should be from people who are qualified to speak about your academic or work competencies and abilities (e.g., your department chair, your manager or supervisor, an instructor that knows your work well). It's okay to include one of your friends as a reference, but make sure you are clear that it is a personal reference. In each case, include the person's name, organization, job title, and relationship to you. Some sample references are provided below.

John Smith, Ph.D. (Grant MacEwan College—Chair, Psychology and Sociology Dept). Dr. Smith is currently my department chair. Phone: (780) 555-4432 Email: smithjohn@grantmacewan.ca

Tom Kurt, Ph.D. (University of Alberta—Professor of Sociology, and Director, Centre for Experimental Sociology). Dr. Kurt was my supervisor for the Honours and Master's program at the University of Alberta. Phone: (780) 555-5485 Email: kurttom@ualberta.ca

Jane Doe, Ph.D. (Grant MacEwan College—Research Council Officer). Dr. Doe is a colleague and friend of mine in the Department of Psychology and Sociology. We both completed the Academy of Leadership Training and Development in 2003. Phone: (780) 555-5768 Email: doejane@grantmacewan.ca

■ OTHER CONSIDERATIONS

How long should your CV be? There isn't a page limit on a CV, but you should try to keep it concise. One way to do this is to replace earlier work with more substantive accomplishments. For example, years from now you might replace your list of essays with abstracts and articles that you published in scholarly journals. An early talk given in a class can be replaced with a teaching appointment. Keep an original CV that has all of your accomplishments in it even if it grows to 10 or more pages. You can always create new, shorter CVs from the original file, depending on the nature of a request. Perhaps you wish to emphasize your teaching experience for a possible position at a college. In this case, you might include information on every presentation, guest lecture, and teaching-related task you have done in the past. Alternatively, if you are applying for a research-based position at a university, you might abbreviate the teaching section but expand on your publication history.

Presentation counts. Print your CV using a laser printer and make any copies using a high-quality copier. Avoid the use of colour and fancy graphics.

Include a cover letter. Usually your CV accompanies a cover letter. The cover letter should be directed to the department head or individual in charge of hiring (e.g., a Director of Human Resources). In a couple of sentences, indicate for what position you are applying (and quote the competition number if applicable), why you are interested in the position, and that you have enclosed a copy of your CV.

What not to include in your CV. While composing the above sections of your CV, try to avoid including any of the following:

- A cute e-mail address in your contact information such as funchick@hotmail.com
- Any interests or hobbies, unless they are directly relevant to the job for which you are applying
- Your religious or political beliefs, marital status, and age
- Reasons for leaving any jobs
- References from relatives
- Long paragraphs—try to use point form whenever possible

- Any negative words or descriptions
- A photo

  Curriculum Vitae Checklist:
- Name and Contact Information
- Education
- Honours and Awards
- Work Experience
  - Teaching
  - Research
  - Professional Presentations
- Publications
- Committees and Associations
- References

REFERENCES

Chapter 1

American Psychological Association (2001). *Publication Manual of the American Psychological Association* (5ᵗʰ ed). Washington: American Psychological Association.

Gollwitzer, P. M. (1999). Implementation Intentions: Strong Effects of Simple Plans. *American Psychologist, 54*(7) 493–503.

Chapter 2

Collins, C. S., & Kneale, P. E. (2001). *Study Skills for Psychology Students: A Practical Guide.* New York: Oxford University Press.

Hirschi, T. (1969). *Causes of Delinquency.* Berkeley: University of California Press.

Durkheim, E. (1867). *Suicide.* (Trans.) John A. Sparkling and George Simpson (1964). New York: Free Press.

Kendall, D., Lothian-Murray, J., & Linden, R. (2004). *Sociology in Our Times* (3ʳᵈ ed.). Toronto: Thomson Nelson.

Chapter 3

Bloom, B. S. (1956). *Taxonomy of Educational Objectives.* (2 vols.) New York: David McKay Co.

Cheser-Jacobs, L., & Chase, C. I. (1992). *Developing and Using Tests Effectively.* New San Francisco: Jossey-Bass Inc., Publishers.

Mentzer, T. L. (1982). Response biases in multiple-choice test item files. *Educational and Psychological Measurement, 42,* 437–448.

Nitko, A. J. (1983). *Educational Tests and Measurement: An Introduction.* Orlando: Harcourt Brace Jovanovich.

Ackermann, E., & Hartman, K. (2000). *The Information Specialist's Guide to Searching and Researching on the Internet and the World Wide Web* (2nd ed.). Chicago: Fitzroy Dearborn Publishers.

Aronson, E., Wilson, T. D., Akert, R. M., & Fehr, B. (2004). *Social Psychology* (2nd ed.). Toronto: Prentice-Hall, Inc.

Berkman, R. (2000). *Find it Fast: How to Uncover Expert Information on Any Subject Online or in Print.* New York: HarperCollins Publishers, Inc.

Blass, T. (1996). Attribution of responsibility and trust in the Milgram obedience experiment. *Journal of Applied Social Psychology, 26,* 1529–1535.

Chan, L. M. (1999). *A Guide to the Library of Congress Classification* (5th ed.). Englewood: Libraries Unlimited, Inc.

Gibaldi, J. (2003). *MLA Handbook for Writers of Research Papers.* New York: The Modern Language Association of America.

Horowitz, L. (1984). *Knowing Where to Look: The Ultimate Guide to Research.* Cincinnati: Writer's Digest Books.

Lane, N., Chisholm, M., & Mateer, C. (2000). *Techniques for Student Research: A Comprehensive Guide to Using the Library.* New York: Neal-Schuman Publishers, Inc.

Mann, T. (1998). *The Oxford Guide to Library Research.* New York: Oxford University Press.

McGuire, M., Stilborne, L., McAdams, M., & Hyatt, L. (2000). *The Internet Handbook for Writers, Researchers, and Journalists.* Toronto: Trifolium Books Inc.

Meuss, W. H., & Raajmakers, Q. A. W. (1995). Obedience in modern society: The Utrecht studies. *Journal of Personality and Social Psychology, 10,* 26–30.

Milgram, S. (1976). Obedience to criminal orders: The compulsion to do evil. In T. Blass (Ed.). *Contemporary Social Psychology: Representative Readings.* (pp. 175–184). Itasca, IL: F. E. Peacock.

Milgram, S. (1974). *Obedience to Authority: An Experimental View.* New York: Harper & Row.

Milgram, S. (1969). *Obedience* [video recording]. University Park, PA: Penn State.

Milgram, S. (1963). Behavioral study of obedience. *Journal of Abnormal and Social Psychology, 67,* 371–378.

Miller, A. G. (1986). *The Obedience Experiments: A Case Study of Controversy in Social Science.* New York: Praeger.

Moscovici, S. (1985). Social influence and conformity. In F. Lindzey & E. Aronson (Eds.). *The Handbook of Social Psychology* (3rd ed.). (Vol. 2, pp. 347–412). New York: Random House.

Reed, J. G. & Baxter, P. M. (2003). *Library Use: Handbook for Psychology* (3rd ed.). Washington: American Psychological Association.

Walker, G., & Janes, J. (1999). *Online Retrieval: A Dialogue of Theory and Practice* (2nd ed.). Englewood: Libraries Unlimited, Inc.

Chapter 5

Akers, R. L. (1991). Self-control as a general theory of crime. *Journal of Quantitative Criminology, 7,* 201–11.

American Psychological Association (2005). *Concise Rules of APA Style: The Official Pocket Style Guide of the American Psychological Association.* Washington: American Psychological Association.

American Psychological Association (2001). *Publication Manual of the American Psychological Association* (5th ed.). Washington: American Psychological Association.

Brownfield, D., & Sorenson, A. M. (1993). Self-control and juvenile delinquency: Theoretical issues and an empirical assessment of selected elements of a general theory of crime. *Deviant Behaviour, 14,* 243–264.

Forde, D. R., & Kennedy, L. W. (1997). Risky lifestyles, routine activities, and the General Theory of Crime. *Justice Quarterly, 14,* 301–331.

Gottfredson, M. R., & Hirschi, T. (1990). *A General Theory of Crime.* Stanford: Stanford University Press.

Grasmick, H. G., Tittle, C. R., Bursik, R. J., & Arneklev, B. J. (1993). Testing the core empirical implications of Gottfredson and Hirschi's General Theory of Crime. *Journal of Research in Crime and Delinquency, 30,* 5–29.

Keane, C., Maxim, P. S., & Teevan, J. J. (1993). Drinking and driving, self-control, and gender: Testing a General Theory of Crime. *Journal of Research in Crime and Delinquency, 30,* 30–46.

Chapter 6

Bard, L. A., Carter, D. L., Cerce, D. D., Knight, R. A., Rosenberg, R., & Schneider, B. (1987). A descriptive study of rapists and child molesters: Developmental, clinical, and criminal characteristics. *Behavioral Sciences and the Law, 5,* 203–220.

Gottfredson, M. R., & Hirschi, T. (1990). *A General Theory of Crime.* Stanford: Stanford University Press.

Rosenkrantz, O. (in press). *Right Your Wrongs.* Toronto: Thomson Nelson.

Symbaluk, D. G. (1997). *An Application of the General Theory of Crime to Sexual Offenders.* Unpublished doctoral dissertation.

Chapter 7

American Psychological Association (1982). *Ethical Principles in the Conduct of Research With Human Participants.* Washington: American Psychological Association.

Sales, B. D., & Folkman, S. (2000). *Ethics in Research With Human Participants.* Washington: American Psychological Association.

Chapter 8

American Psychological Association (1982). *Ethical Principles in the Conduct of Research With Human Participants.* Washington: American Psychological Association.

Sales, B. D., & Folkman, S. (2000). *Ethics in Research With Human Participants.* Washington: American Psychological Association.

APPENDIX 1: SAMPLE ESSAY

Heaven's Gate: A Multitheory Analysis

Submitted by:
Student ID:
Submitted to:
Course:
Date Submitted:

In March 1997, in a large house in Rancho Santa Fe, California, 39 members of a group called "Heaven's Gate" made their "Final Exit" from this life. It was neither murder nor suicide; it was an exit (Sager, 1997). This "Exit" was considered a highly anticipated event for the group, and 39 people participated not only voluntarily but also joyfully. However, for the rest of society, this event was regarded as possibly nothing more than a cult of brainwashed individuals who were coerced into taking their own lives. This pejorative interpretation must be understood within the social context of a small yet recent history in our society of homicidal/suicidal religious groups: Order of the Solar Temple, Hare Krishnas, Branch Davidians, Aum Shinri Kyo, and Jonestown. Yet it is also important to understand the difference between these groups and Heaven's Gate: the members of Heaven's Gate premeditated an elaborate suicide and were not forced (neither internally nor externally) to participate, but rather complied voluntarily. Although it must be stated that it is impossible to verify whether or not coercion was used on that specific occasion, it is evident that years of coercive tactics and a "top-notch indoctrination program" played a crucial role on members' decision to participate in the "Exit" (Geier, 1998: 32).

Heaven's Gate is a complex phenomenon that cannot be understood using a simple level of analysis. There are various aspects of this group that require not only initial but further research, and as a result, data regarding this new religious movement is extremely limited, consisting mostly of newspaper and magazine articles. It is the intention of this paper to theoretically analyze three of the most crucial aspects of Heaven's Gate's existence and eventual demise: how members joined, why they stayed, and why they committed suicide. By detailing a brief history of the group and using a combination of functional, social psychological, and criminological theories, the complexity of this group, the need for further research and consequently new interpretations will become evident.

Heaven's Gate was an ecclesiastical religious group that combined "elements of Buddhism, Hinduism, Christianity, Astrology" and belief in UFOs (Sager, 1997: 392). The founders of this new religious movement, Marshall Herff Applewhite and Bonnie Lu Trusdale Nettles (a.k.a. Guinea and Pig, Bo and Peep, and finally Do and Ti), met in 1972 in Houston, Texas (Sager, 1997). Applewhite had recently been through a series of life-altering events, including a divorce, a dismissal from his teaching position at the University of Saint Thomas in Houston after having a relationship with a male student, and the death of his father, a Presbyterian minister. Applewhite became so devastated, distraught and full of sexual loathing, he committed himself for psychiatric treatment (Sager, 1997). With much in common, Applewhite and Nettles became inseparable almost immediately. They gradually withdrew from all friends and family and continued to deepen the strong bond they shared. Convinced they had been sent here on a mission from God, they began to travel in an attempt to fulfill their duties (Sager, 1997).

Upon finding their destinies, Applewhite and Nettles taught of a Kingdom of Heaven ruled by a group of gods who directed that humans be seeded in a type of cultivation experiment that was carried out on earth. Periodically, "Representatives" from the Kingdom came to earth to deposit souls into their test subjects in preparation for transference to the "Evolutionary Level Above Human." Harvest time occurs approximately every 2,000 years and consists of a Representative taking a human body in order to harvest all souls deposited and transport them back to the "Next Level." Applewhite claimed to be the successor of Jesus as the assigned Representative from the Kingdom (Sager, 1997: 391–92). Applewhite also taught of his and Nettles' mission to fulfill the prophecy set forth in the Book of Revelations, 11:3–13. This passage tells of two prophets who are killed and then rise from the dead and ascend to the Kingdom of Heaven in a cloud of light (believed to be a UFO), triggering Armageddon and preparing earth for the next seeding (Sager, 1997). Applewhite described the "Process" to his followers as a path of asceticism: the renunciation of all material possessions and all ties to society, "[m]an must break from all ties that bind him to this earth: mother, father, brother, sister" (Applewhite as cited in Sager, 1997: 392).

According to Mike Sager of GQ magazine, Applewhite and Nettles, the "Older Members," employed many types of control mechanisms over the group to ensure the Process. All younger members of the group were stripped of their given names, and each member was given a similar name of three letters, no vowels and the suffix –*ody* (derivative of God) to signify their immaturity. Applewhite and Nettles prescribed a strict daily regimen and rules that were to dictate every aspect of daily life and were to be followed to the letter. All members wore the same clothing (an oversized men's shirt and unisex slacks), they wore bedroom slippers and their heads were shaved in an effort to remove all signs of gender and eliminate all distinctions. They were assigned seating, sleeping arrangements, duties, diets and hygienic practices (hygiene alone was detailed in three pages of the Procedure Manual). Members' only possession was a toothbrush and all the money (the group's monthly income was $40,000) was given to the "Pursers," who kept detailed accounts of every penny. All members were required to be partnered, yet idle conversation was not permitted. Any type of sexuality was absolutely forbidden; therefore, celibacy was held as the golden rule. Applewhite had himself castrated, he claimed, in an effort to quell his sexual impulses from interfering with his spirituality. Following their leader, seven other members also had themselves castrated in what they believed was a logical extension of celibacy. All forms of weakness (e.g. vanity, gender, curiosity, friendliness, etc.) were not only to be conquered but also had to be recorded so they could be shared with the rest of the "Class" at the weekly "Slippage Meeting." Slippages were also recorded in an "Eyes Log," as it was considered an offence to trust your own judgment (Sager, 1997).

As a result of strict mechanisms of social control, members began to rely upon the group for everything and became completely dependent on others, in turn losing any sense of autonomy. Members were directed how to act, how to think and feel, therefore resulting in total compliance to group ideologies. Members were forced to live the most private aspects of their lives publicly. They were forced to cut all conventional ties to their family, adopt a new name, live in a place where bedrooms were communal, bathrooms were public, and public confessions were a weekly event and were regarded as a measure of devotion (Sager, 1997).

As compliance to rules was an essential condition of group membership, unbelievably, new members could still be recruited. One might ask why someone would choose to join a restrictive, dictatorial group such as Heaven's Gate. The answer to this lies with Edwin H. Sutherland, a pioneer of criminological thought, who developed a theory of differential association that could be used to explain religious conversion. Originally, Sutherland's theory was used to explain criminal and deviant behaviour. However, it is a social learning theory that Sutherland notes "attempts to explain crime via learning, interaction and communication" (Martin, Mutchnick, & Austin, 1990: 147).

Sutherland combined macro-level and micro-level theories in an attempt to explain both individual and group behaviour, specifically the "social and mental elements of the learning process" (p. 149). Although differential association has been used to explain criminal behaviour, it is important not to disregard such a critical social psychological approach to understanding individual learning processes in a group context. Criminal conversion and religious conversion have very similar criteria, and therefore Sutherland's differential association may be used to explain how Heaven's Gate was able to attract and recruit new members.

Sutherland's theory of differential association is a series of nine propositions detailing how criminal behaviour is learned. In order for this theory to be applied to religious conversion, the term "group behaviour" will be substituted for the term "criminal behaviour" and the term "societal norms" will be substituted for the terms "legal codes" and "laws."

Sutherland begins his theory with the proposition "Group behaviour is learned" (Martin, Mutchnick, & Austin, 1990: 156). The second proposition follows that "Group behaviour is learned in interaction with other persons in a process of communication" (p. 156). Sutherland continues to propose that "The principle part of the learning of group behaviour occurs within intimate personal groups" (p. 156). This third proposition is emphasized in the Heaven's Gate requirement that recruits travel in groups of two and that all tasks be performed with a partner. Members of Heaven's Gate were rarely left to themselves because when individuals are with others, they put on a public performance of self and tend not to question or doubt assumptions in an environment of assurance (Roberts, Hollifield, & McCarty, 1998).

Sutherland's fourth proposition is that "When group behaviour is learned, the learning includes (a) techniques of the group; (b) the specific direction of motives, drives, rationalizations and attitudes" (Martin, Mutchnick, & Austin, 1990: 156). In the context of religious conversion, this point is exemplified in Heaven's Gate's implementation of a phone-book-sized "Procedure Manual" and the rationalization that "only the most committed would survive" and be able to "find fulfillment beyond their wildest dreams in the Next Level" (Sager, 1997: 394, 399). The group's dictatorial rules and regulations include members having to chew every mouthful 24 times before swallowing and not being permitted to have private thoughts (Sager, 1997). Moreover, the promise of a better life to the most devout (often equated with compliance) provided members with motive and drive to follow such a strict life. The group rationalizations of this life of total obe-dience consisted of statements such as "all were equal in the eyes of the Next Level" in relation to "the interchangeable nature of the soul deposits," as well as not allowing "the vehicle to become habituated to anything earthly" (p. 392, 394); because there is nothing earthly about the Next Level. The attitude of the members was one of sacrifice for a higher goal. The mem-bers had been engrained with the belief that "[b]eing hard on yourself was a sign of devotion" (p. 399). Therefore, devotion came to imply commit-ment which would yield "circumstances 10,000 times more fulfilling than anything on the Human Level" (p. 387–88).

Sutherland's fifth proposition is that "The specific direction of motives and drives is learned from definitions of societal norms as favourable or unfavourable" (Martin, Mutchnick, & Austin, 1990: 157). Heaven's Gate pro-posed that compliance to societal norms would not achieve transplantation to the Next Level. It was necessary to break all ties and not develop attachments to anything earthly if members wanted to go to the Next Level. Members of Heaven's Gate did not necessarily view societal norms as unfavourable. Rather, they may have been viewed as not conducive to achieving the goal of the group.

Sutherland's sixth proposition is that "A person becomes a member of the group because of an excess of definitions favourable to violation of societal norms" (Martin, Mutchnick, & Austin, 1990: 157). In the case of Heaven's Gate, there was an unbelievably long list of rules and regulations that violated societal norms as well as governed every aspect of daily life for members. For example, in an experiment done years ago, members were required to wear black hoods at all time in an effort to eliminate personal interaction between members that could effect their spiritual attainment (Sager, 1997).

Sutherland's seventh proposition is that "Differential associations may vary in frequency, duration, priority and intensity" (Martin, Mutchnick, & Austin, 1990: 157). This statement may be the key to determining religious conversion. It implies that the more individuals associate with a group and the more intense the experience, the more likely a conversion will take place. When this is done in conjunction with limiting associations with non-members, it is almost certain the individual will convert. This differential association is exemplified in Heaven's Gate's demands for members to break all ties with family and friends. Members were permitted to associate only with other members, which is evident in the group's need to partner all members.

Sutherland's eighth proposition is that "The process of learning group behaviour by association with members and non-members involves all mechanisms that are involved in any other learning" (Martin, Mutchnick,

& Austin, 1990: 157). Techniques of learning include reading, memorization and application. Therefore, learning to become a member of a religious movement is done in the same manner as learning to become a functional member of society. Heaven's Gate possibly used this theory of learning when they implemented a "phone-book-thick legal-size loose leaf Procedure Manual," that members were told to read when they had nothing to do (Sager, 1997: 394). Heaven's Gate did not tolerate errors when it came to the Manual and forced its members to learn appropriate group behaviour.

Sutherland's ninth and final proposition is that while group behaviour is an expression of general needs and values, it is not explained by those general needs and values, since non-group behaviour is an expression of those same needs and values (Martin, Mutchnick, & Austin, 1990). This last proposition poses difficulties in the context of religious conversion. In sociological terms, the functional definition of religion is the provision of meaning and order. Therefore, when individuals become vulnerable to conversion, it is due to a lack of meaning and order in their lives, which is eventually fulfilled by a religious movement. It is plausible that individuals who do not become involved in a religion are provided with meaning and order in another sphere of life and are forced to seek the satisfaction of these needs elsewhere. Therefore, although religion's provision of meaning and order is a sufficient explanation for conversion, it is not necessarily the only explanation. Once members are converted and committed to a religious group, the question arises as to why individuals would remain in a restrictive religious movement. The motives and drives members experience that compel them to remain in a group appear to be quite questionable to non-members. In the case of Heaven's Gate, it seems strange that so many members would remain in such a restrictive group for long periods of time. At the time of the group's demise, some members had been involved since its conception in 1975 (Sager, 1997). In order to explain why individuals do not leave religious groups, it is necessary to employ the social psychological theory of deindividuation.

Deindividuation refers to "the loss of a person's sense of individuality and the reduction of normal constraints against deviant behavior" (SP, 1999: 256). Groups fundamentally undermine people's individual experiences, upon which they base their moral foundations with externally imposed cues. Deindividuation occurs as a result of a combination of predispositional factors and situational contingencies, such as anonymity, arousal and reduced feelings of individual responsibility (SP, 1999). In order for deindividuation to occur, the group must normalize taboos. In our society, sex serves as the source of procreation, a source of entertainment, and a form of intimacy. It is a taboo in our society to choose celibacy over sex. Celibacy is reserved for the devoutly religious while sex is common practice in our society. In the context of Heaven's Gate, all forms of personal interaction and physical contact were limited to a bare minimum, and celibacy was viewed as the golden rule. Furthermore, when the leader of Heaven's Gate had himself castrated, seven other members followed (Roberts, Hollifield, & McCarty, 1998). The castrations were performed in an effort to achieve spiritual purity. In our society, castration is perceived as genital mutilation.

The second criterion of deindividuation is to provide people with new reference groups. Heaven's Gate fulfilled this by providing members with a relatively isolated environment, resulting in a reference group comprised solely of group members.

The third criterion for deindividuation is for members to break from their past. This includes receiving a new name, dissociating from family members and friends, and reinterpreting past experiences. In relation to groups renaming members, Heaven's Gate provided all members with similar names, each made up of three letters, no vowels and the suffix –*ody*. The names were given to signify members' immaturity but at the same time, removed each individuals sense of identity. In the case of individuals dissociating from their pasts, members were urged to break all ties with family and friends that bound them to this earth (Sager, 1997). In the context of members reinterpreting past experiences, Heaven's Gate implemented a weekly Slippage Meeting as a means of forced confessions. Members were required to document all offences and share them with the class on a weekly basis (Sager, 1997). These Slippage Meetings became a sign of devotion for members, and as a result, members began to create list of offences ("the worse the better") to unveil at the weekly meeting (p. 392).

The fourth criterion of deindividuation is for the group to establish non-normative social roles. Heaven's Gate achieved this by following rules of asceticism, "the belief that the renunciation of physical pleasure will enable [people] to be released from the body to find union with the divine" (Ferris et al., 1997: 32). In our society, the accumulation of wealth is considered to be a top priority and deserving of respect. Therefore, a group of people who willingly renounce all material possessions for the attainment of a higher goal are considered to be deviant in terms of normative social roles.

The fifth and final criterion for deindividuation is for members to hold no secrets. The implementation of a no-secrets rule helps groups to foster vulnerability in individual members. Members are forced to make confessions that are then replaced with group doctrines and ideologies. In the case of Heaven's Gate, a cult of confession was established where "shame and harsh judgment [were] used to ensure the psychological vulnerability of cult members" (Roberts, Hollifield & McCarty, 1990: 7). It is similar to the reinterpretation of past experiences, where members were forced to make confessions at a weekly Slippage Meeting. As well, Heaven's Gate used an "Eyes Log" to document offences committed by members. These were also shared at meetings because members could not be trusted to reveal all offences. Furthermore, members' secrets were replaced with group doctrines and ideologies contained in the Procedure Manual. The manual dictated proper thoughts and behaviours as well as a long, comprehensive list of violations. Both mental and behavioural procedures had to be followed completely in order for transplantation to the Evolutionary Level Above Human to occur (Sager, 1997).

This process of transplantation to the Next Level became the sole purpose for Heaven's Gate's existence. Since the group's conception in 1975, the goal was evident: avoid Armageddon, reach the Next Level, and prepare to have all of your wildest dreams fulfilled. The process to achieve this was viewed as strict yet feasible. The only potential problem in the Process was the most important step. The "Final Exit," death, was the definitive moment of the Process. It was at this time that the ultimate compliance was required but not coerced. It was a decision made voluntarily by every individual member, yet the decision was made based on promises which would come to fruition in an unverifiable realm. The decision to take their lives was made solely based on faith. It is at this point the question arises as to why individuals would choose to commit suicide. The sociological answer can be found in Emile Durkheim's study of *Suicide*.

In 1897, Durkheim released one of his most famous studies, *Suicide*, in which he created a useful typology of suicide: egotistic, altruistic, anomic and fatalistic (Ritzer, 1992). Durkheim found that integration into society and regulation by society were the key determinants of suicide. For our purposes, it is only the egoistic and altruistic types of suicide that will be discussed. On the one hand, egoistic suicide is characterized by low degrees of integration, "the individual is not well integrated into the large social unit" (Durkheim, reprinted in Ritzer, 1992: 90). According to Durkheim, altruistic suicide is characterized by high degrees of social integration. In the case of the latter, the degree of integration is so strong that the individual is actually coerced, either actively or passively, into committing suicide (Ritzer, 1992).

In the case of Heaven's Gate, years of manipulation, passive coercion and faithful compliance affected members' ability to make consensual decisions in regards to the Exit. The founders, Applewhite and Nettles, made grandiose promises without providing members with any tangible evidence to verify their claims. Applewhite described the Evolutionary Level Above Human as inexplicable. He knew that only "[t]here was no gender, no conversation, no individuality, no sex [and] no food" (Sager, 1997: 387). Life on the Next Level was incomprehensible. "It was a physical place, but its exact nature escaped human words and concepts" (p. 387).

Implementation of the Procedure Manual was done to help members adjust when the transplantation occurred. Members were told that the strict regulations of the Manual were needed and enforced on earth as a means of preparing them for their duties on the Next Level, where "the fates of galaxies would be in [their] hands," and it was critical "to get things exactly right" (Sager, 1997: 394). Whilst making promises of a more fulfilling life, they were also threatening the ultimate punishment: Armageddon. A critical aspect of Heaven's Gate theology was that *after the Final Exit* had been executed, all those who remained on earth would succumb to Armageddon. Given that the members were forced to choose between a horrible death in Armageddon or a painless death that would lead to a better life, it is no wonder members chose the latter. The probability of altruistic suicide "springs from hope, for it depends on the belief in beautiful perspectives beyond this life" (Durkheim reprinted as cited in Ritzer, 1992: 90).

The interpretation of Heaven's Gate being nothing more than a brainwashed cult forced to committing suicide now becomes more complex when analyzed in conjunction with a multitude of theories. It is limiting to use only traditional theories to explain and interpret religious group behaviour; further sociological perspectives are not only relevant but also essential for a fuller understanding of religious conversion. Given the rise of homicidal/suicidal religious groups at the dawn of this millennium, further research and new interpretations are needed to give insight into past and present religious groups, as well as possibly prevent further incidents from occurring. As we continue to educate ourselves on the cult/counter-cult debate, new solutions may arise that lead to new and more effective means of intervention. This begets the question: whose responsibility is it to intervene? Only through education can intervention occur and lead to the prevention of future tragedies.

References

Brehm, S. S., Kassin, S. M., & Fein, S. (1999). *Social Psychology* (4th ed.). Boston: Houston Mifflin Company.

Ferris, T., Chauncey, G., Hubbard, L. R., Ross, A., Schiff, S., & Gladwell, M. (1997, April 14). De-programming Heaven's Gate. *The New Yorker, 73*(8), 31–33.

Gerier, T. (1998, March 30). Is there life after death for Heaven's Gate? A year after mass suicides, the cult carries on. *U.S. News & World Report, 124*(12), 32.

Martin, R., Mutchnick, R. J., & Austin, W. T. (1990). *Criminological Thought: Pioneers Past and Present.* New York: Macmillan Publishing Company.

McGuire, M. B. (1997). *Religion: The Social Context* (4th ed.). Belmont, CA: Wadsworth Publishing Company.

Ritzer, G. (1992). *Sociological Theory* (3rd ed.). New York: McGraw-Hill, Inc.

Roberts, L. W., Hollifield, M., & McCarty, T. (1998). Psychiatric evaluation of a "monk" requesting castration: A patient's fable with morals. *American Journal of Psychiatry, 155*(3), 415–420.

Sager, M. (1997, September). Stairway to heaven. *Gentlemen's Quarterly, 67*(9), 384–400.

APPENDIX 2: Sample Research Report

Running Head: Impression Formation

Key Words: Impression formation, central traits, warm cold, person perception.

The Warm-Cold Variable in Forming First Impressions of Personality

Lisa Moldaver

Grant MacEwan College, Edmonton, Alberta

Date submitted: April 6, 2001

*Research was supported by a student loan in the name of the primary author and a substantial donation from Mary Stan Galbraith. Correspondence and request for reprints can be directed to Lisa Moldaver, Department of Social Sciences, Grant MacEwan College, 10700 – 104 Avenue, Edmonton, AB, T5J 4S2.

Abstract

The present study was undertaken in an effort to duplicate Asch (1946) and Kelley's (1950) classic findings that central personality traits "warm" and "cold" influence impression formation. In a study of twenty-two college students, participants were given background information about a person and then were exposed to a tape recording of that person undergoing a Thematic Apperception Test. Impressions were given using a semantic differential scale and descriptive paragraphs written by the subjects. Consistent with Asch and Kelley's early research, results of this study showed that subjects exposed to the warm condition formed much more favourable impressions of a stimulus person than subjects in the cold condition.

Introduction

Forming impressions is something we engage in every day. First impressions are the basis for determining the like or dislike we feel towards others. In a job interview, blind date, or bank loan application, we are conscious of the fact that others will be judging us by the impression they form. A positive impression can bring benefits whereas a negative impression can produce harmful outcomes (e.g., failure to get the job or being declined for the bank loan). Therefore, it is in people's best interests to portray themselves in a positive manner and to be described by others in a positive way.

The subject of how people perceive others is an important question to social scientists. It has been studied for over 50 years, beginning with Solomon Asch. In 1946, Asch conducted several experiments concerning impression formation, using Gordon Allport's (1937) classification of personality traits. Allport distinguished among personality traits using the categories cardinal, central and secondary. A cardinal trait was defined as one trait that dominates almost all of the person's personality; very few people possess this type of trait. A central trait was described as the foundation of personality. Most people have less than 10 central traits, and these traits are very influential in the general dispositions of people. A secondary trait was based on situations and therefore the last in the hierarchy of personality.

Using Allport's early classification, Asch (1946) conducted experiments with central and secondary (peripheral) traits. He read subjects a list of "character qualities" including intelligent, skillful, industrious, determined, practical and cautious. He differentiated the list by using either the term "warm" or "cold." One group heard the list with the word *warm* while the other group heard the list with the word *cold*. Asch's subjects were then asked to write a brief description of the person described by the traits. Participants also completed a checklist of opposing traits, selecting the trait they felt best fit their description. Asch (1946) found that inclusion of the central trait *warm* or *cold* produces very different impressions, with *warm* creating a much more positive impression relative to *cold*. Asch also noted that peripheral traits (e.g., polite and blunt) have little or no effect on impression formation.

In another experiment, where identical lists of traits were read to one group of subjects then reversed in order when read to a second group, Asch discovered a primacy effect whereby the first words guided the formation of an impression in a specific direction (either positively or negatively). McKelvie (1991), however, was unable to replicate Asch's primacy effect with a different list of traits. These findings support previous claims that the primacy effect cannot be generalized to other word lists. It remains unknown what feature of the original list produces this effect.

In 1950, Harold Kelley replicated the procedures of Asch's original experiment with one exception. Kelley tested whether the same results could be reached if participants were exposed to an actual person following the background information. A confederate posed as a teacher, and subjects were asked to write a description of the stimulus person and also fill out a fifteen-item rating scale. Kelley found the same distinct consistencies between the warm group and the cold group as Asch had found in 1946. Participants exposed to warm background information gave consistently more favourable impressions, participated in class more, and interacted more than those who received cold prior information about the teacher.

The main objective of the current study is to examine the effects of warm/cold background information on impression formation more than fifty years after the original studies. Asch found that impressions could be formed and generalized by reading a list of traits of a particular person without ever introducing the stimulus person to the experiment. Kelley expanded these results to find that central qualities also affect impressions of real people. The purpose of the current study is to determine the effect of central traits in background information about a stimulus person (who is heard on a pre-recorded tape) on the general impressions people form, using a controlled environment and manipulating the central trait: warm/cold.

Method

Participants

Twenty-three Grant MacEwan College students in a second year experimental psychology course agreed to participate in this study. The procedure was conducted during class time and with the instructor's express permission.

Setting and Materials

The regular classroom for the experimental psychology course at Grant MacEwan College was used for the location for this experiment. Materials included a script, two different background information papers (one including the word "warm," one including the word "cold"), a semantic differential scale, a tape player and a prerecorded cassette of two confederate voices.

Procedure

Before the commencement of the experiment, a script for the researcher was constructed in order to have a controlled environment (See Appendix A). The participating students were informed that they would hear a tape of a prerecorded session between a patient and a psychologist, during which a Thematic Apperception Test is administered to the client. Participants were then instructed to assess the client in order to compare their impressions with that of the psychologist following the study. After this procedure, participants were asked to sit so that everyone was spaced apart in an effort to avoid influencing others' assessments (in actuality, however, this was to ensure none of the subjects knew of the different background information sheets that were given to them, in order to preserve the validity of the manipulation of the independent variable). Subjects were also asked not to comment while the tape was playing.

Thematic Apperception Test

The researcher then described the Thematic Apperception Test (TAT) which was developed in 1935 by Morgan and Murray of the Harvard Psychological Clinic. Psychologists use this to assess personalities and aid in personality research. It employs a series of vague pictures, which the client is then instructed to create corresponding stories that they believe describes the events depicted in the photographs. The pictures are deliberately broad and general so that the patient places him- or herself into the stories constructed. The researcher in this study noted that several scoring and interpretation systems have been developed in reference to the TAT as well as the fact that the TAT has had many criticisms brought against it in recent years. Subjects were told of one of the major problems involving the dated quality of the pictures which may influence interpretations and also limit the overall reliability and validity of the test. The researcher informed the subjects that most often, the TAT is used to make quantitative and qualitative analyses of the client's story formation in order to assess aspects of the client's personality that has been revealed through the application of the TAT.

Description of Tape Recording

The experimenter explained to the subjects that they would hear a tape-recorded session of a TAT application. The researcher also noted that the tape is very quiet, so participants should listen carefully, because after hearing the session, they would be asked to write a descriptive paragraph (i.e., a qualitative analysis) about the client on the tape. It was again reinforced that subjects should not talk amongst themselves or comment until the exercise was complete, though subjects were encouraged to write down impressions formed while listening to the tape. Subjects were then given a sheet of background information relating to the client on the tape and asked to read it through before listening to the taped TAT session. There were two different background information sheets, and each participant received one of the versions. The descriptions were identical with the exception of only one word (see Appendix A). (Prior to distribution, the background sheets were shuffled to ensure random distribution among the subjects.) One sheet gave a description of the client as a *cold*, practical, determined and industrious person whereas the other background sheet gave a description of the client as a *warm*, practical, determined and industrious person. After participants read the background information, they heard the tape recording of the TAT application.

Quantitative and Qualitative Analyses of Stimulus Person

Following the tape, subjects were asked to write a descriptive paragraph about the client on the tape (the qualitative analysis) on the back of the information sheet. The semantic differential scale was then handed out and subjects were asked to complete the scale to provide some quantitative feedback. The scale consisted of 20 bipolar adjectives (opposing personality traits). Subjects were instructed to circle the number closest to whichever trait the client's personality consisted of. It was explained that the 7 coincides with the trait on the right and the 1 coincides with the trait on the left. The farther towards any one end of the scale they went, the stronger they felt the trait represented or did not represent the client's personality, and the middle number was more indicative of a neutral impact. Subjects were instructed to read each trait carefully because the positive and negative traits were not listed in the same order, meaning a positive trait could be either on the left or the right side of the scale, and likewise with the negative traits. The alternating format of the scale ensured that participants didn't simply use a set response pattern in answering that would affect the validity of the research. Subjects were then given as much time as needed to complete both qualitative and quantitative analyses.

Manipulation Check

For the manipulation check, subjects were asked, after completing the analyses, to write on the back of their paper what they believed the research hypothesis was that would have been tested based on what they had just completed.

Debriefing

Researchers then debriefed subjects by explaining that they had just unknowingly participated in a classic experiment developed by Solomon Asch. They were informed that the true purpose of the study was to examine the effect of background information on impression formation. Specifically, this study looked at how the information they received affected their assessment of the person undergoing the Thematic Apperception Test. It was explained that all aspects of the experiment (instructions, tape and forms) were identical except for the independent variable (i.e., the background information), which was randomly distributed among them so that some people received information containing the descriptor "warm" while others received information that included the trait "cold." The researcher noted that deception was necessary to preserve the study's validity. If the true purpose had been known, the analyses given would be different. Participants were invited to contact the researcher if they had any questions or concerns regarding any aspect of the study.

Early Research

Following the debriefing, researchers described the early research by Solomon Asch (1946) and Harold Kelley (1950), which was the basis for this study. It was then explained that this study was designed to find out if the same results could be duplicated 54 years later. Subjects were also told of the specific hypothesis that differences in personality ratings and descriptions are predicted to be a function of receiving either the word "cold" or "warm" as part of the background information. Subjects were then asked if some volunteers in both the warm and the cold conditions could read their statements along with some descriptor words to illustrate the effects. Researchers followed this by reading descriptions from previous demonstrations of the warm-cold study.

Scale Inversion

Next, the experimenter explained that the semantic differential scale was created with alternating items to prevent a set response. As a result, some bipolar traits needed to be inverted so that the positive trait always corresponded to the highest number (7) and the negative trait always corresponded to the lowest number (1). Researchers gave examples of how to do this (e.g. sociable 1 **2** 3 4 5 6 7 after the scale inversion would become sociable 7 **6** 5 4 3 2 1) and then confirmed which traits needed to be inverted. Each participant then inverted his or her scale for subsequent data analysis.

Data Collection

Researchers collected the semantic differential scales and written descriptions from subjects. T-tests were conducted to compare the means for participants in the "warm" and "cold" groups. In addition, measures of central tendency were calculated for both conditions along with correlations that imply the degree and strength of the relationship between variables. To complete the manipulation check, subjects' hypotheses were examined, and data pertaining to anyone who hypothesized the true nature of the experiment was omitted from the study.

Dependent Variable—General Impressions

General impressions of the stimulus person were measured as ratings on twenty personality traits using a seven-point semantic differential scale. Items on the scale were inverted so that the lowest score (1) corresponded to the negative aspect and the highest score (7) corresponded to the most positive aspect of each trait. Researchers coded the seven-point scale as follows:

1 = Strongly disagree with positive trait

2 = Moderately disagree with positive trait

3 = Disagree with positive trait

4 = Neutral

5 = Agree with positive trait

6 = Moderately agree with positive trait

7 = Strongly agree with positive trait

Results

Upon completion of the manipulation check, it was found that one subject in the cold condition (C8) had correctly hypothesized the true nature of the experiment and therefore that person's impressions and semantic differential scale were omitted from the analysis.

Quantitative Analyses

Of the twenty traits on the scale, nine were inverted so the highest possible score per trait always corresponded to the positive adjective, allowing reliable statistics to be calculated (see Appendix B). When the total scores of the semantic differential scale were summed individually, participants' scores were evenly distributed between conditions. Eleven of the scores appeared in the negative level, none in the neutral level and eleven in the positive level (see Table 1).

Table 1 about here

The null hypothesis was expected to be that the difference between the overall mean for warmth by people given the warm condition subtracted from the overall mean of warmth for people given the cold assessment would be zero.

$$H0 = Xsdep1 - Xsdep2 = 0$$

The alternative hypothesis was that the difference between the mean assessments of warmth in the warm and cold conditions would be less than zero.

$$H1 = Xsdep1 - Xsdep2 < 0$$

An independent samples t-test with 22 subjects was conducted using 20 (i.e., n-2) degrees of freedom and the conventional level of significance of 0.05. The test statistic (the number of standard errors that separate the observed difference from the expected value of zero) was 2.792 at the 0.004 significance level (the probability of getting the observed difference) for a one-tail (directional) test, t(22) = 2.79, p = <.05. Therefore, the decision was made to reject the null hypothesis (see Table 2).

Table 2 about here

The mean for each personality trait was calculated for both conditions (see Table 3). The warm group had an overall mean of 104.72 compared to 84.00 in the cold condition, with a resulting difference of –20.73 (see Table 3).

Table 3 about here

The modal mean score for the bipolar traits in the warm condition was a rating of 5 indicating overall support for the positive side of the scale. Mean ratings in the cold condition tended to be 4's and 3's indicating neutrality or disagreement with positive bipolar traits.

Qualitative Analyses

Most people formed generally positive impressions in the warm condition and generally negative ones in the cold condition. The following are some examples from the warm condition:

> *The woman ... is imaginative, and seems good at inferring emotions ... she seems to be a gentle, conservative woman.*

> *She is cooperative, caring and sensitive ... concerned about the well being, comfort, and happiness of others.*

In the cold condition, many subjects used the words *conventional, traditional, conservative, apathetic,* and *unimaginative* in their descriptions of the stimulus person. One subject described the person as "cold, matter of fact" and who "doesn't seem to care about anything." These descriptions lend further support to the hypothesis that exposure to either the word "warm" or "cold" results in an impression that generalizes beyond the personality description for the stimulus person. This is especially evident in descriptions that extend beyond personality such as "traditional family values" and a "happy, stable family life ... [being] cooperative, sensitive and caring" based solely on the central trait of *warm* or *cold*.

Discussion

This study supports the findings of the classic research on impression formation. Participants exposed to the central trait *warm* in background information of a stimulus person did form much more favourable impressions than those exposed to cold background information. Ratings on the individual personality traits were generally higher in the warm compared to cold condition. Results of the qualitative analyses indicated that participants in the warm condition attributed all kinds of favourable characteristics to the stimulus person, including many traits that extend beyond the individual's personality. Alternatively, those who received negative background information tended to rate the stimulus person very unfavourably.

The fact that the means for the cold condition were largely negative and neutral scores may reflect the influence of another variable that was not controlled for in this study (e.g., gender, socioeconomic status or education level of the participant). It could also reflect a central tendency bias that was a result of people not wanting to choose an obviously negative or positive trait (i.e., it is much less judgmental to be a fence-sitter). Similarly, presentation of the stimulus person may also affect the outcome of the experiment. The gender of the stimulus person should ideally be alternated and monitored during future replications of the experiment. The manner used to deliver the information regarding the stimulus person (e.g., video, tape-recording, etc.,) may also differentially influence the impact of warm/cold traits in the formation of impression and should be examined in more detail.

In conclusion, impression formation was affected by prior background information, and the results obtained substantiate the previous claims made by Solomon Asch (1946) and Harold Kelley (1950). Impression formation and impression management are extremely important activities in everyday life. They can bring beneficial (e.g., receiving the bank loan to buy your first home) or harmful effects (e.g., failing to get the job needed to repay the bank loan). It is important to understand how people perceive others in order to create and maintain positive impressions that may be extended to all aspects of personality. When meeting people for the first time, it is important to be mindful of how we portray ourselves and how we are described by others, because creation of a negative impression could have damaging consequences, whereas creation of a positive impression may result in substantial rewards.

References

Asch, S. E. (1946). Forming impressions of personality. *Journal of Abnormal Social Psychology, 41*, 258–290.

Kelley, H. H. (1950). The warm-cold variable in the first impressions of persons. *Journal of Personality, 18*, 431–439.

McKelvie, S. J. (1991). Effect of design on the Asch primacy effect. *Journal of Social Psychology, 131*(5), 751–753.

Table 1. Overall Impression by Condition

Impression	Warm Condition	Cold Condition	Total
	n	n	n
Negative	2	9	11
Neutral	0	0	0
Positive	9	2	11

Table 2. Results of Independent Samples T-test of Scale Means for Each Condition

Valid N of Cases	Degrees of Freedom	Test Statistic	P-Value	One or Two Tail Test
22	20	2.792	.008/2 = .004	One

P-value	Relationship	Level of Significance	Decision
.004	<	.05	Reject null hypothesis

Table 3. Means for Individual Personality Traits by Condition

Personality Trait	Mean for Warm Group	Mean for Cold Group
Sociable	4.7273	3.2727
Self-Assured	4.8182	4.5455
Intelligent	5.9091	5.4545
Good-natured	5.1818	3.9091
Interesting	4.0909	3.1818
Attractive	5.0000	3.8182
Tolerant	5.3636	3.9091
Sincere	6.2727	4.9091
Kind	5.9091	3.9091
Courteous	5.9091	5.0909
Warm	5.2727	2.9091
Unselfish	5.6364	4.0909
Wise	5.5455	4.6364
Honest	5.9091	4.9091
Gentle	6.0909	5.1818
Motivated	5.0909	5.1818
Witty	3.6364	2.9091
Pleasant	5.5455	3.8182
Strong	4.7273	4.3636
Active	4.0909	4.0000
Scale Total	104.73	84.00